Global Warming, Environmental Governance and Sustainability Issues

Global Warming, Environmental Governance and Sustainability Issues

Editor

Will McConnell

MDPI • Basel • Beijing • Wuhan • Barcelona • Belgrade • Manchester • Tokyo • Cluj • Tianjin

Editor
Will McConnell
Woodbury University
USA

Editorial Office
MDPI
St. Alban-Anlage 66
4052 Basel, Switzerland

This is a reprint of articles from the Special Issue published online in the open access journal *Sustainability* (ISSN 2071-1050) (available at: https://www.mdpi.com/journal/sustainability/special_issues/Environmental_Governance_Sustainability).

For citation purposes, cite each article independently as indicated on the article page online and as indicated below:

LastName, A.A.; LastName, B.B.; LastName, C.C. Article Title. *Journal Name* **Year**, *Article Number*, Page Range.

ISBN 978-3-03943-212-7 (Hbk)
ISBN 978-3-03943-213-4 (PDF)

Cover image courtesy of Will McConnell.

Contents

About the Editor

Will McConnell received his Ph.D. from McMaster University, where he began combining research modes from the sciences, social sciences, and humanities. He has published and delivered mixed-mode research internationally across several disciplines. His research designs solutions in multiple areas of ocean and land-based research; these research areas combine the chemistry of the ocean and land-based cycles of consumption, ocean and atmospheric pollution, and the production and perception of waste; social marketing and community-based social marketing applied to changing human behavior and the understanding and awareness of agricultural production techniques and ocean issues; political discourse and best practices policy and decision-making in combining environmental, economic, and social forms of sustainability; aridity, the production of GHGs and bio-agriculture of the south-west United States region; food desertification and the production and distribution of food in the United States; international law and policy of non-territorial ocean areas; successful reef protection, management, and regeneration efforts globally, with a focus on marine protected area (MPA) strategies; terrorism, the state, and theories of revolution and revolt. His most recent work focuses on the emerging issues of global ocean deoxygenation and the design of solutions for better tracking of oxygenation levels across the major upwelling current systems of the global ocean (eastern boundary upwelling systems—EBUS). He is an avid scuba diver, having incorporated diving into the study of many underwater regions across the globe; while on land, he has an abiding interest in xeriscaping and zero waste urban and architectural design. He is now based in Los Angeles, California.

Preface to "Global Warming, Environmental Governance and Sustainability Issues"

Many thanks to Woodbury University for supporting interdisciplinary research across the years I have been conducting research there as a member of the faculty. I would also like to thank my faculty colleagues there and across the globe who are engaged in innovative research design to address the many issues of sustainability—especially Dr. Douglas Cremer, Dr. Eric Schockman, Dr. Rossen Ventzislavov, Dr. Emily Bills, and Professor Mike Sonksen, whose support and conversations always fueled my curiosity. Many thanks to Lisa Cooper and Ofelia Huidor, whose friendship and calm, steady presence always keeps everyone around them heading in the right direction. A special thanks to Kiki, for always believing in me, somehow, no matter how daunting a project seemed.

Will McConnell
Editor

Editorial

Introduction to Sustainability Journal Special Edition "Global Warming and Sustainability Issues"

Will McConnell

Interdisciplinary Studies, Woodbury University, Burbank, CA 91504, USA; Will.McConnell@woodbury.edu

Received: 7 July 2020; Accepted: 9 July 2020; Published: 15 July 2020

Sustainability, in its multiple facets, is nothing if not interdisciplinary. In research circles, the current challenges to defining an alternative balance among global financial, economic, and environmental sustainabilities in global environments now produce almost constant reconsiderations, even upheavals, in what and how we can be said "to know"—including how to discern our best strategies for achieving sustainability. In this charged atmosphere, the boundaries between and among academic disciplines constantly undergo recontextualizing gestures in the search for more refined research paradigms, paradigms whose research metrics increasingly need to be constructed from across disciplinary research methods, models, and versions of validity. This same feature of current research methods is also a driving agent for the re-definition and re-inscription of disciplines driven by more specialized, more discipline-specific forms of modeling in academic institutions. Across our discursive sphere, we seem to need to have it both ways: on the one hand, we need disciplinary forms of research in order to enable more precise interdisciplinary modeling; on the other hand, we need to push disciplinary research methods into forms of overt and often less-than-predictable, sometimes underestimated and understated, collusion in order to break through the limitations of the more discrete, disciplinary development and application of models.

The essays included in this special edition attempt to place these tensions into new and emerging contextual "workings" within and against one another—among financial/economic determinants and environmental imperatives; across science, social science, and humanities disciplinary boundaries; and in reconfiguring qualitative and quantitative research methodologies and findings, to propose new directions for the constellation of financial/economic, political/social, and human/environmental patterns for achieving sustainability, if, by "sustainability", we mean the horizon of intelligibility bequeathed that word by the *Brundtland Report* (*Our Common Future*) [1]. The essays in the current edition of *Sustainability* all take the *Brundtland Report* as a starting point only, however; whereas the authors of the Brundtland Report never intended to develop knowledge that has specific applications in delimited social, geographic, and socio-economic policy recommendations and contexts, the authors of the current essays do take the difficult steps to integrate the wider vision of the Brundtland Report with the (often) context-specific application of their findings in their regional situatedness.

Like many strong collections that appear as a result of specific, multiple pressures, emerging as moments in the history of developing and/or emerging concepts and practices, this collection of essays provides a sample of developing strategies in agricultural processes and practices; identifies needs for the local and global development of modeling and paradigms; and, broadly speaking, construes findings that could benefit many regions across the globe. Similarly, these studies offer multiple combinations of disciplinary assumptions, research strategies, and assumptions of validity. Taken together, ultimately, this collection produces a cross-pollination of models and emerging research directions that represent secure steps for governments revising their research approaches to create more resilient conditions for their citizens as for their ability to "sustain" specified levels of living conditions for their current (and projected) populations. Across the globe, we are only in the beginning stages of the deliberate application of many of the scientific findings evident in the collected essays; that is, the authors here identify specific problems in the current design of research approaches, and these essays rebuild knowledge as they contribute research-based findings to address shortcomings or oversights in the relationship between social policy, living conditions and situations, and scientific data. Thus, these essays gesture toward a Brundtlandian world in which scientifically based findings are integrated directly into policy considerations and design, and these considerations, in turn, are reflected in the ultimate needs of the people and planet. The studies in this special edition of *Sustainability*, then, gesture toward an overall approach that not only calls for an awareness of our impacts on the planet and our own resilience in a future life increasingly driven, and limited, by Green House Gas (GHG) emissions but will also, one day, protect as it builds upon current levels of environmental resources, the overall environmental health of ecosystems and, ultimately, through combinations of the research approaches on display here, foster rather than diminish the health of the planet.

Although the essays in the collection range across the globe in terms of applicability and insight, the majority of the studies included here begin by developing geographic specificity in the application of a scientific method to an existing or emerging problem that is at once social and atmospheric, economic and environmental, driven by both planetary and human systems. Although few of the authors explicitly pursue the destabilizing of existing models in climate change science, the assumptions here are governed by the need to develop models with what I will call "geographic transferability": wider applicability globally in *specific* geographies and social systems, for increased accuracy or predictability in the ability to lower emissions to safer levels for all humans. The current challenge of this set of goals, of course, is to shift to low-carbon economies and modes of production without sacrificing human food and basic living necessities or securities; this necessity has emerged as the key challenge of our epoch. Each of the researchers gathered in this special edition understands that this challenge demands the redesign of climate impact modeling, both across the globe and regionally, in a strategy of geographic localization. The problem is, now, to develop the discussion of what these models actually mean in a world organized, or reorganized, by equality of access to resources rather than by a hinterland–metropolis power struggle for survival in conditions of dwindling resources in conditions of rapid climate warming.

For example, in "Mitigation of CO_2 and N_2O Emission from Cabbage Fields", Hwang, Park et al. [2] pose the possibility of developing a strategy of mitigation through paying closer attention to the interaction of CO_2 and N_2O emissions via monitoring tillage depth and nitrogen levels in cabbage fields in Korea. They attempt to alter the current processes for strategic, socio-political decision-making in order to "incentivize the shift towards sustainable farming". By developing the configuration of data differently—producing, through their study, advances in modeling—their insights lead quietly, if quickly, to a different configuration of "private" and "public" organization in society; they investigate a more sophisticated approach to developing low-carbon practices in farming, ultimately proposing the development of a gradated system of compensation for farmers who, in adopting low-carbon tillage practices, may thereby reduce their overall cabbage yields. While the emphasis of the study is, rightly, on developing more sophisticated modeling of the relationships among tillage depth, nitrogen fixation in the soil, and GHG emissions as a result of different levels adopted in the relationship between tillage depth and nitrogen fertilization, the study's tentative conclusion gestures toward a revised social and agricultural system in South Korea's (near) future. What these agricultural-social policies might become, however, is largely dependent on any given nation's willingness to redesign, based on emerging scientific modeling, long-standing social, political, and economic assumptions, commitments, and paradigms.

The authors Glab and Sowiński [3], also investigating effective GHG mitigation strategies, turn their attention to the sustainable production of sweet sorghum as a bioenergy crop; however, they develop a model of the carbon footprint of sorghum production through the use of sewage sludge and digestate as a nutrient substitute. Their study takes place within the geopolitically "localized" context of the European Union but develops an additional mitigation opportunity that could further reduce the GHG impact of currently carbon-intensive agricultural practices. Thus, read these two studies together, applying the findings across the continuities of their localized parameters, suggests that the modeling strategies these authors' work may be combined most effectively into are as follows. A single strategy combining the findings of the two essays has the potential to reduce the percentage of subsidies proposed by Hwang and Park's study as necessary to meet GHG emissions goals—provided that these goals are defined clearly and, furthermore, that the GHG emissions goals are linked, through advanced modeling, to specific agricultural processes and practices that are measurable, attainable, and modifiable in atmospheric and other, more socio-politically determined, conditions in the future. This approach, exciting as it is, requires considerable additional study, the analysis of findings across sustained study parameters, and the development of resources as well as confidence in the methodological paradigm shift that both studies develop. As Glab and Sowiński note [3], the basic building blocks for such an approach require development: "little data are available on GHG emissions from sweet sorghum production under temperate climate. Similarly, information on the effect of bio-based waste products use on the carbon (C) footprint of sorghum cultivation is rare in the literature". The need to refocus analysis on the role of carbon—as well as the role of different methods for applying fertilizers in different locations and across differing farming practices—is also crucial to bear in mind, as Glab and Sowiński make clear: "Nitrogen application had the greatest impact on the external GHG emissions and it was responsible for 54% of these emissions".

In another exciting study in this special edition, Yang, Long et al. [4] attempt to develop models that map the environmental cost of a "typical" citrus-producing county in China. They find, too, that the "production and utilization" of nitrogen fertilizer is a significant object of study: nitrogen fertilizer "accounted for more than 95% of the total environmental costs" of citrus production in the areas that formed the basis for the study. Their findings are stunning: based on 155 farmers' data, "the high yield and nitrogen use efficiency orchard group with younger and better educated owners, achieved a higher citrus yield and N use fertilizer efficiency with less fertilizer input and lower environmental costs" (1). The intervention these authors suggest sounds practical, relatively simple to implement, and achievable across diverse cultural, political, and scientific contexts globally. We would do well to pause on this finding and attempt to replicate the authors' study methods, inroads to more sophisticated modeling, and overall findings in other areas of the globe: as the authors note, "citrus is the top fruit crop with the largest cultivation area and highest production in the world". The object of study, the relationship between citrus yields and productivity assumptions in the current, widely adopted regimes of fertilizer application, is remarkably strategic, not merely for China's adaptability and resilience in the face of global climate change but for all nations attempting to provide for the current generations' needs without compromising the ability of future generations to meet their needs. As the authors note, "optimum nutrient management based on the local field recommendation in the citrus-producing areas is crucial for achieving a win-win target of productivity and environmental sustainability in China and other, similar countries".

While their work points to exciting insights that lie just beyond our own more nation-based decision-making paradigms, the study moves toward developing models that provide the capacity to extrapolate across multiple environmental, social, and political conditions the findings expressed there. However, due to the relative lack of development, in the current conditions for producing knowledge, for producing widely accepted pairings of scientifically-based knowledge, the development of specifically interdisciplinary-informed models, with socio-political policy and decision-making, the authors' inroads into alternative modeling methodologies can compel only a tantalizing horizon of possibilities for future study, suggest the development of an emerging global awareness of interdisciplinary modeling, and hint at a paradigm of equality—of more balanced access, stability, and security, for all. As they note, "although many aspects of environmental costs in cereal and annual crop production have already been investigated, the life cycle assessment (LCA) of environmental indicators in perennial fruit crops is still rare, mainly due to lack of methodological standardization". In part, this limitation in the studies is driven by significant differences across geographic regions: soil types, climate conditions, management practices, language, and other cultural barriers, etc. However, "methodological standardization" should not be the "crucial" determining factor in our inability to make inroads to sustainability. To eliminate this false barrier to change, the quantification of environmental sustainability continues to be a singular challenge.

The movement toward a more aggrandizing perspective is made by Wu, Huang et al. in "Net Greenhouse Gas Emissions from Agriculture in China" [5]. The authors of the study take "21 sources of agricultural GHG emissions into consideration", linking the "emission" and "absorption" of GHGs in a single model—in this case, focusing on emissions in 30 provinces in China between 2007 and 2016. The study analyzes the spatial correlation and convergence of net GHG emissions in China's agricultural production. They find that, in the "agricultural GHG emission structure of China" across that period, as we might expect, there are fluctuating patterns of net emissions across provinces; however, across their findings, the rate of "absorption was much lower than emissions". Thus, their findings are multiple, but one key finding is that, in the results of the convergence tests of 30 provinces' GHG emission scenarios across the period under study, there is no nationwide convergence, which suggests that emissions will not decline by natural means. The development of effective reduction measures is a crucial finding of the study, and furthermore, the data showed that strong possibilities exist for regional cooperation in the sharing of low-carbon technologies. For its future, China should "attach importance to the development of the technologies and techniques" of low-carbon technologies (and non-technological strategies), applying these "as soon as possible". Furthermore, the authors find that "at present, low-carbon technology is relatively insufficient for agriculture in China". These findings diverge from current studies conducted in China, in that the authors widen the modeling inputs from 6 to 21 sources of GHG emissions. This, of course, is a significant rethinking in the design of GHG modeling in agricultural applications, and we would do well outside of China to pattern additional studies based on the model developed by these researchers.

Many of these themes, challenges, and inroads into modeling on global scales are developed, albeit with a different emphasis, by Jensen, Domínguez et al. ("Economic Impacts of a Low Carbon Economy on Global Architecture: The Bumpy Road to Paris") [6]. Taking a similar pathway in arguing for increased accuracy in modeling a, or the, global carbon economy, the authors call for an "integrated modeling framework". The study gathers considerable data from across multiple existing disciplinary frameworks; for example, "first, the macroeconomic impacts of moving into a global low carbon economy are analyzed" by applying different carbon taxes in a general equilibrium modeling framework". Emission mitigation technologies are then quantified and applied through the Aglink–Cosimo model to assess agricultural markets' responses compatible with emissions scenarios mirroring the 2 °C threshold of the Paris agreement. While the authors underscore the need to create substantial reductions in GHG emissions and foster the transition to a climate-friendly, low-carbon economy, their findings also express the need for caution in the development of a successful set of GHG mitigation strategies. As they word their conclusion, "transition to a lower carbon intensive economy has large implications from both regional and global perspectives". An environmental modeling of successful strategies or clear policies must take into account, equally, economic, environmental, and societal impacts. Only then will the transmission of these strategies directly into (local and global) agricultural market policy be "fair" in assessing the long-, medium-, and short-term effects of changes in the global agricultural market systems. Their analysis indicates that "for the net mitigation of global agricultural emissions", policies need to target carefully specified interventions in current policy and practice decisions. Policy and other considerations for emerging geospatial emission intensities, or addressing measurable "hotspots" of CO_2 production in localized areas, differ between developing and developed countries, for example. More sophisticated and differentiated policy approaches are needed in the agricultural sector, and this, in turn, could be driven more effectively by movement to an updated version of the Aglink–Cosimo model, measuring agricultural productivity in numerous regions of the globe, the dimensions of which the authors develop and extend to readers of the work. In the end, the revised modeling scenario developed by the authors can fuel technological development in correlating carbon tax scenarios to more differentiated, context-specific environmental, societal, and political conditions—suggesting not only the interdisciplinarity of the authors' research and modeling but also the need for additional, similar studies able to tackle the design of a methodology capable of a global scale of "localizable" applicability.

The final two essays in the edition address aspects of the increasingly arid conditions in many regions, brought about by emerging and predicted climate change patterns in the precipitation and aspiration of rainfall due to increasing aridity in many regions across the globe. Like the other studies in the collection, these authors situate their work in the latest predictions emerging from sources such as the Intergovernmental Panel on Climate Change (IPCC) as well as other experts, whose thinking has formed a largely accepted set of framework ideas for addressing the mitigation of, adaptation to, and reversal of the detrimental impacts of climate change on human (and other) populations. As Yi Li, Xie et al. note [7], "climate change has altered the existing pattern of precipitation and has an impact on the resistance and adaptability of desert plants". Studying life on the antipodes of human adaptability is an emerging strategy for testing the limit cases of adaptability and mitigation in life forms already adapted to the stress of what, for humans, are largely inhospitable climatic conditions.

The authors position their work in an emerging paradigm of climate change study: as climate change increases dramatically, impacts on precipitation are expected to alter landscapes differentially, if simultaneously, across multiple regions globally: "the time and intensity of precipitation may change". Here, too, the study of specific climatic changes is a key feature of the study parameters the authors construct: although in the semi-arid and arid regions of northwestern China, precipitation is projected to increase from 30 to 100 mm in the next 100 years, this increase will be accompanied by a "trend of increasing precipitation intervals, decreasing small precipitation events and increasing extreme precipitation events". The character of precipitation events across many additional areas of the globe is expected to undergo these forms of geo-spatial differentiation—a diffusion of current climate into regional "aridities" characterized by vastly different precipitation events than those we see in current climatology. Climatologists expect flash flooding accompanied, paradoxically, by increasing aridity over time. The authors underscore the reach of their study; although the authors limit their object of study to a specific region and set of climatic conditions in northern China, "globally, the proportion of land surface under extreme drought is predicted to increase from 1–3% currently to 30% by the end of 2090". Underscoring the urgency of studies of arid conditions is the fact that this percentage of change in arid regions globally is likely to be underestimated in current modeling scenarios, simply because the complexity of the data continues to be a challenge to the development of an adequate model; in the absence of clear, more comprehensive collection and analysis of data for arid regions, models of this kind of climate change have long remained on the conservative side in the predication of projections across longer temporal spans. At stake is a more comprehensive knowledge of impacts we might expect—or be able to mitigate more effectively—in the destabilization of ecological successions of species forced to adapt beyond current morphological processes and, more rapidly, to conditions of drought and less frequent, more intense patterns of precipitation. As the authors note, the interactive impacts, as well as the central characteristics of the changes in precipitation amounts and frequency in desert ecological conditions, are "unclear" (1). The study is an attempt to contribute baseline data for additional study in an area of knowledge production currently not well known or understood.

As the authors characterize their contributions, given that the main change in precipitation is in the "intensity, frequency, and duration" of precipitation events, more research is necessary for predictive and mitigation efforts across these specific variables; similarly, few scholars compare "frequency and interactive impacts on individual plant growth". Thus, their study highlights the interactive properties of rainfall by placing into contiguity, for example, precipitation amount and precipitation frequency. Their findings, as well as their study design, have significant implications for developing more knowledge of emerging semi-arid and intensified arid regions' conditions for sustaining climate, creating refuge conditions or areas in semi-arid and arid regions, and the development of collection and dispersal methods for these emerging patterns of rainfall in regions long perceived to be "wastelands" or un- or under-productive geo-climatic resources. The authors unarticulated insight, that studying existing semi-arid and arid regions through simulated rainfall and frequency conditions will accrue predictive and adaptive value, is directly applicable now. Shifting the perception of arid regions now can produce a more informed, longer-term set of strategies for building human resilience in areas predicted to become arid with rapid changes in climate in this century.

Kamali, Abbaspour et al.'s work [8]—"A Quantitative Analysis of Socio-Economic Determinants Influencing Crop Drought Vulnerability in Sub-Saharan Africa"—in some senses, reverses the parameters of the above study, although these two studies, taken together, offer a glimpse of the potential for exacerbated aridity posed by unabated atmospheric conditions in the emergent patterns of climate change. Population studies have long predicted that the world population will increase significantly by 2100, with estimates often predicting an increase of 2 billion people (from 7 billion to 9 billion) in that relatively brief time span. Pressures on food and water security are expected to increase significantly across the century, with the greatest vulnerabilities expected in South Asia and in Sub-Saharan Africa. Combining social science research modes with scientific data, the authors assert, rightly, that the social infrastructures in these regions not only produce significant impacts on levels of human adaptability but also may contribute negatively to drought resilience. The study combines the use of the crop drought vulnerability index (CDVI), the application of data from the Environmental Policy Integrated Climate model (EPIC), and the identification and quantification of socio-economic variables in regressive analysis techniques.

Their results circle back to where we began the collection of essays: "the level of fertilizer use" is a highly influential factor in understanding, and measuring, vulnerability. For most of us, of course, fertilizer is the "forgotten" element of food production, environmental sustainability, and climate change. However, as multiple studies in the collection of essays point out, "fertilizer" is the "forgotten element" that may turn out to have the most significant impacts on land, ocean, and atmospheric health, simply because this is a key, readily available, easily understood point of widescale, global intervention. Understanding more precisely the interaction of these systems of human assumptions, agricultural processes, or practices, and the earth's responsiveness to changes in both, is a fundamental pathway to resilience, mitigation, and, in many cases, survival, as climate change's impacts become increasingly difficult to avoid. The unvoiced hope of these studies, of this collection's contributions to sustainability *as* adaptability, is that the interdisciplinarity and analytic rigor of the work of these scholars contributes, quite literally, not only to a better future but also to a more secure set of pathways to ensure greater equality of access to survival itself.

References

1. Report of the World Commission on Environment and Development: Our Common Future. Available online: https://sustainabledevelopment.un.org/content/documents/5987our-common-future.pdf (accessed on 23 February 2020).
2. Hwang, W.; Park, M.; Cho, K. Mitigation of CO_2 and N_2O Emission from Cabbage Fields in Korea by Optimizing Tillage Depth and N-Fertilizer Level: DNDC Model Simulation under RCP 8.5 Scenario. *Sustainability* **2019**, *11*, 6158. [CrossRef]
3. Glab, L.; Sowiński, J. Sustainable Production of Sweet Sorghum ass a Bioenergy Crop using Biosolids Taking into Account Greenhouse Gas Emissions. *Sustainability* **2019**, *11*, 3033. [CrossRef]
4. Yang, M.; Long, Q.; Li, W. Mapping the Environmental Cost of a Typical Cirtus-Producing County in China: Hotspot and Optimization. *Sustainability* **2020**, *12*, 1827.
5. Wu, H.; Huang, H.; Tang, J. Net Greenhouse Gas Emissions from Agriculture in China: Estimation, Spatial Correlation and Convergence. *Sustainability* **2019**, *11*, 4817. [CrossRef]
6. Jensen, H.; Dominguez, I.; Fellmann, T. Economic Impacts of a Low Carbon Economy on Global Architecture: The Bumpy Road to Paris. *Sustainability* **2019**, *11*, 2349. [CrossRef]

Sustainability **2020**, *12*, 5671

7. Xie, Y.; Li, Y.; Xie, T. Impact of Artificially Simulated Precipitation Patterns Change on the Growth and Morphology of Reaumuria soongarica Seedlings in Hexi Corridor of China. *Sustainability* **2020**, *12*, 2439. [CrossRef]

8. Kamali, B.; Abbaspour, K.; Wehrli, B. A Quantitative Analysis of Socio-Economic Determinants Influencing Crop Drought Vulnerability in Sub-Saharan Africa. *Sustainability* **2019**, *11*, 6135. [CrossRef]

Article

Mitigation of CO_2 and N_2O Emission from Cabbage Fields in Korea by Optimizing Tillage Depth and N-Fertilizer Level: DNDC Model Simulation under RCP 8.5 Scenario

Wonjae Hwang, Minseok Park, Kijong Cho, Jeong-Gyu Kim and Seunghun Hyun *

Department of Environmental Science and Ecological Engineering, Korea University, Seoul 02841, Korea;
hwj0145@korea.ac.kr (W.H.); asithinkyou@korea.ac.kr (M.P.); kjcho@korea.ac.kr (K.C.);
lemonkim@korea.ac.kr (J.-G.K.)
* Correspondence: soilhyun@korea.ac.kr; Tel.: +82-2-3290-3068

Received: 19 September 2019; Accepted: 31 October 2019; Published: 4 November 2019

Abstract: In this study, we applied the Denitrification and Decomposition model to predict the greenhouse gas (GHGs; CO_2 and N_2O) emissions and cabbage yields from 8072 cabbage fields in Korea in the 2020s and 2090s. Model outputs were evaluated as a function of tillage depth (T1, T2, and T3 for 10, 20, and 30 cm) and fertilizer level (F1, F2, and F3 for 100, 200, and 400 kg N ha^{-1}) under the Representative Concentration Pathways 8.5 climate change scenario. For both time periods, CO_2 emissions increased with tillage depth, and N_2O emissions were predominantly influenced by the level of applied N-fertilizers. Both cabbage yields and GHGs fluxes were highest when the T3F3 farming practice was applied. Under current conventional farming practices (T1F3), cabbage yield was projected at 64.5 t ha^{-1} in the 2020s, which was close in magnitude to the predicted cabbage demand. In the 2090s, the predicted cabbage supply by the same practice far exceeded the projected demand at 28.9 t ha^{-1}. Cabbage supply and demand were balanced and GHGs emissions reduced by 19.6% in the 2090s when 94% of the total cabbage farms adopted low carbon-farming practices (e.g., reducing fertilizer level). Our results demonstrate the large potential for Korean cabbage farms to significantly contribute towards the mitigation of GHGs emissions through the adoption of low-carbon farming practices. However, in order to incentivize the shift towards sustainable farming, we advise that lower yield and potential economic losses in farmlands from adopting low-carbon practices should be appropriately compensated by institutional policy.

Keywords: climate change; greenhouse gas; cabbage farming; DNDC model

1. Introduction

According to the Intergovernmental Panel on Climatic Change (IPCC) [1], the global atmospheric concentration of carbon dioxide has increased from pre-industrial levels of ~280 ppm to 391 ppm in 2011 due to greenhouse gas (GHGs) emissions from the industrial, transportation, and agricultural sectors. Agricultural cropland accounts for approximately 11% of the global land area and 4.8% of the total global GHGs emissions [2,3]. In particular, upland farming is considered a significant source of carbon dioxide (CO_2) and nitrous oxide (N_2O) to the atmosphere and is responsible for 26% of the total GHGs emissions in Korea's agricultural sector [4,5].

The Representative Concentration Pathways (RCP) adopted by the IPCC assumes that the continued and rapid increase of GHGs emissions will cause a projected global mean temperature rise of 1.0–3.7 °C by the late 21st century. Rising temperatures and altered precipitation patterns associated with global climate change are expected to further aggravate agricultural GHG emissions due to their influence on the carbon (C) and nitrogen (N) dynamics of cultivated soils [6,7]. It is therefore necessary to improve agriculture resilience to climate change through climate-adapted farming practices in order to minimize GHGs emissions.

A wide variety of crops, such as cereals (e.g., corn, wheat, and barley) and vegetables (e.g., cabbage and radish), are cultivated in upland fields in Korea [8]. Cabbage in particular is an essential vegetable in Korea, as it is a main ingredient of kimchi: a traditional side dish of salted and fermented vegetables [8,9]. Over the past few decades, farmers have applied agronomic practices, such as fertilization, irrigation, and tillage to improve crop yield. Tillage practices in upland farming can improve soil aeration through soil loosening, while the application of fertilizers promotes crop yield through the addition of essential elements [10,11]. The Korea Rural Development Administration (KRDA) has therefore established optimal fertilizer application and tillage depth recommendation levels for various agricultural crops. For cabbage cultivation, a standardized limit of 400 kg ha^{-1} of N fertilizers and a tillage depth of 10 cm are recommended.

The challenge of modern agriculture in response to climate change is to simultaneously improve crop yield and reduce GHG emissions. Managing both tillage depth and fertilizer application can influence GHG emissions due to their direct impacts on soil organic carbon (SOC) oxidation and the activity of soil denitrifying bacteria [11,12]. The majority of research to date has focused on the impacts of agronomic practices on either crop yield or GHG emissions individually under the different climate change RCP scenarios [6,7,13]. However, it is necessary to consider the simultaneous impacts of climate change on both crop production and GHG reduction to further our understanding on sustainable agriculture in response to climate change.

The denitrification and decomposition (DNDC) model is particularly advantageous, as it can estimate crop yield and GHGs (e.g., CO_2 and N_2O) emissions simultaneously [14]. The model can also simulate the overall impacts of various agricultural practices on model outcomes [15]. The DNDC model predicts crop yield by employing the plant growth algorithm from the Crop Environment Resource Synthesis (CERES) model, which has been validated in Korea using long-term national statistical data and field measurements [16,17]. Further, GHG emissions are assessed based on biochemical reactions of soil microorganisms in response to the immediate meteorological conditions [18].

Policies to reduce GHGs emissions are being developed and implemented worldwide. Policies such as the "Carbon Farming Initiative" of Australia and the "Grassland Ecological Incentive and Subsidy Policy" of China provide incentives to farmers through the control of agronomic practices [19]. Similarly, the Korean government plans to implement its "low-carbon farming" policy, which is a direct payment program that will provide compensation to farmers who voluntarily engage in GHG reduction through the adoption of agronomic practices [20]. The extent of GHGs reduction from participating farms should therefore be adequately assessed to determine the appropriate level of compensation for low-carbon farming practices. Arable lands in Korea are typically small in size and are distributed sporadically throughout the country [9]. According to the administrative record on agricultural products, Korea harbors 8072 cabbage farms with field sizes ranging between 0.01–94 ha. Large uncertainties would therefore arise in model predictions using large spatial scales due to inaccuracies associated with scattered and smaller-sized farmlands. It is therefore necessary to include high-resolution data to accurately predict and quantify the extent of GHG reduction achieved by adopting low-carbon practices.

In this study, we applied the DNDC model to predict both the level of GHGs emissions and crop yield in South Korean cabbage fields in response to climate change (RCP 8.5 scenario) and under different farming practices. We generated high-resolution (1 km^2) weather and soil data for the model input by re-synthesizing raw data (12.5 km^2) obtained from the Korea Institute of Public Administration. For farming practices, we included nine sets of input variables by combining three N-fertilizer application levels (100, 200, and 400 kg N ha^{-1}) and three tillage depths (10, 20, and 30 cm). The Korean cabbage fields were split into 8072 1 km^2 grid cells. The objectives of this study were to (i) validate the DNDC model's capability to predict GHGs (CO_2 and N_2O) emissions from cabbage farms based on field measurements under different tillage depths and levels of applied N-fertilizer;

(ii) simulate GHGs emissions and crop yield in Korea as a function of tillage depth and the level of fertilizer applied; and (iii) identify best cabbage farming practices to maximize cabbage production, minimize GHGs emission, and balance the future supply and demand of cabbage under the RCP 8.5 scenario.

2. Methods and Materials

2.1. Model Prediction

We applied the DNDC model (version 9.5; http://www.dndc.sr.unh.edu/) to predict the annual average cabbage yield (t ha^{-1} yr^{-1}) and GHGs emissions (t CO_2-eq ha^{-1} yr^{-1}) in the 2020s and 2090s from Korean cabbage fields under the RCP 8.5 scenario. The DNDC model is a process-based biogeochemical model with a temporal resolution of one day and is based on the circulation of carbon and nitrogen in agricultural ecosystems [14]. Both CO_2 and N_2O emissions were converted to CO_2-equivalents (CO_2-eq) using the reported values of specific global warming potentials [1].

2.1.1. Study Sites

The geographical location and distribution of cabbage fields in South Korea are present in Figure S1a. South Korea is located between 33°09′ and 38°45′ N, and 124°54′ and 131°06′ E with a total national land area of 100,019 km^2. We extracted the cabbage cultivation land cover data (1:25000 scale) from the Korea Ministry of Environment (KME) by incorporating the KRDA's land suitability guideline for cabbage cultivation. The total cabbage field area was ~3200 km^2, and cabbage farmland contributed 11.1–17.8% of the total upland area in the period of 1980–2015 [9]; the observed percentage variability is due to variability in previous year cabbage market prices.

2.1.2. Model Input Data

The model input data (i.e., climate, soil, and farming practices) are listed in Table 1. For climate data, we used the RCP 8.5 climate change scenario based on the Hadley Centre Global Environmental Model Version 3 Regional Atmosphere (HadGEM3RA) [21]. RCP 8.5 is a scenario of comparatively high greenhouse gas emissions and corresponds to the highest degree of projected climate change. We obtained daily temperature and precipitation data from 2011 to 2095 at a 1 km^2 scale from the Korea Meteorological Administration (KMA, http://climate.go.kr). For soil data, we obtained initial SOC and pH from the KRDA's soil database of approximately 365000 datapoints. Clay content and bulk density were reprocessed at a spatial resolution of 1 km^2 for 377 domestic representative soil series based on a 1:25000 scale soil map.

We combined three tillage depths (10, 20, and 30 cm for T1, T2, and T3, respectively) and three N-fertilizer levels (100, 200, and 400 kg N ha^{-1} for F1, F2, and F3, respectively), equivalent to a total of nine farming practice input variables (e.g., T1F1, T1F2, T1F3, T2F1, T2F2, T2F3, T3F1, T3F2, and T3T3), to assess the effects of farming practices on model outcomes. We applied the KRDA method in the model simulation to include other input data such as cultivation period, timing of fertilization, and plowing, etc. For cabbage farming in Korea, a tillage depth of 10 cm (T1) and a N-fertilizer application of 400 kg N ha^{-1} (F3) are recommended by the KRDA. This farming practice combination (T1F3) is therefore referred to as the "conventional method" in this study.

We used the average climate and soil data from 2006 to 2015 as the model baseline for the 2016 to 2095 model-run period. We conducted the 80-year simulation from 2016 to 2095 following the 5-year 2011 to 2015 run-up. The results for each ten-year period (2016–2025 and 2086–2095) were averaged and referred to as the 2020s and 2090s, respectively.

Table 1. Details of the input data for the model predictions, field verification, and baseline evaluation.

Data Type	Sub-Type	Unit	Model Prediction		Field Verification [f]
			Future Scenario [d]	Model Baseline [e]	
Climate [a]	Temperature	°C	RCP 8.5	Mean values between 2006 and 2015	Administrative data for Deokso field in 2018
	Precipitation	cm			
Farming Practice [b]	Fertilizer level	kg N ha^{-1}	F1, F2, and F3	F3 (conventional method)	F1 and F3
	Tillage depth	cm	T1, T2, and T3	T1 (conventional method)	T1 and T3
Soil [c]	Bulk density	g cm^{-3}	Administrative soil database	Administrative soil database	Field measurement data
	Clay	%			
	Initial SOC	g kg^{-1}			
	pH	1:5			

a. Both the Representative Concentration Pathway (RCP) 8.5 and administrative climate records were obtained from the Korea Meteorological Administration. b. Fertilizer levels F1, F2, and F3 denote 100, 200, and 400 kg N ha^{-1}, respectively. Tillage depths T1, T2, and T3 denote 10, 20, and 30 cm, respectively. The period of cultivation was designated between August 11 and October 31 for all farming practices. c. Soil data from the Korea Rural Development Administration (KRDA) was used for model prediction. SOC is the soil organic carbon. d. The DNDC prediction was performed for the 2020s and 2090s under RCP 8.5 and as a function of the different farming practices. e. For the baseline evaluation (i.e., no future climate change), average climate data of the last 10 years (2006–2015) and the KRDA method (10 cm tillage depth and 400 kg N ha^{-1} fertilizer) were used as input parameters. f. Model verification was performed at the Deokso field. Soil properties were measured on-site.

2.1.3. Identifying Best Farming Practices to Achieve Three Scenario Goals

In this study, we aimed to identify the best farming practices (tillage depth and fertilizer level) to achieve the following scenarios: (1) minimum GHGs emissions, (2) maximum cabbage production, and (3) appropriate levels of cabbage production to balance future cabbage demand. To accomplish these goals, we ran the model by assigning one of the nine farming practices (e.g., T1F1, T1F2, T1F3, T2F1, T2F2, T2F3, T3F1, T3F2, and T3T3) to each of the 8072 grid cells until the desired goal was achieved. We began by assigning T1F3 (i.e., the conventional method) to all grid cells for comparison. We obtained model outputs for annual average cabbage yield (t ha^{-1} yr^{-1}) and GHGs emissions (t CO$_2$-eq ha^{-1} yr^{-1}) for all cases.

In order to identify the scenarios of lowest national GHG emissions (scenario 1), we input low-carbon farming practices to the grid cells in which the GHG emissions from conventional practices (T1F3) were greater than the national average. To identify the scenario of highest cabbage production (scenario 2), higher-yield farming practices were applied to the grid cells in which cabbage yield were lower than the national average. For the third scenario model run, we forecasted the future cabbage demand in the 2020s and 2090s using the Korea agricultural simulation model (KASM) developed by the Korea Rural Economic Institute [22]. Demand forecasting was performed based on the analysis of past cabbage demand and trade under present market conditions. To therefore identify the scenario resulting in crop yield closest to contemporary demand, we applied low-carbon farming practices to the grid cells in which conventional farming cabbage yields were greater than the demand.

2.2. Field Measurements

2.2.1. Experimental Site and Data Collection

To verify the DNDC model prediction capabilities, we measured in-situ GHGs emissions under varied farming practices (tillage depth and fertilizer level) in a cabbage field (37°35′01″ N, 127°14′16″ E) located at the Korea University Agricultural Farm (Deokso field) in Gyeonggi Province. A combination of upper and lower limits for tillage depth (T1 and T3) and fertilizer level (F1 and F3) were selected for field validation, which therefore includes T1F1, T1F3, T3F1, and T3F3 (Table 1). We measured CO$_2$ and N$_2$O emissions approximately once every month from March to November 2018 (9 samples for CO$_2$ and N$_2$O emissions for each set). The soil properties of the Deokso field were as follows: bulk density of 1.2 ± 0.1 g cm^{-3}, clay content of 22 ± 2.3%, initial SOC content of 3.0 ± 0.2%, and pH of 6.2 ± 0.1 (1:5 with H$_2$O).

2.2.2. Measurements of CO_2 and N_2O Emissions

We conducted the one-year GHGs measurements using the closed chamber method modified from our previous work [23]. The system consists of a closed chamber and a measurement unit in which a moisture filter, direct current (DC) pump, flow meter, CO_2 detector module, and data logger are sequentially connected (Figure S1b). Gas sampler ports with a cock valve were installed on top of the chamber. An opaque acrylic cylinder chamber with a volume of 25.1 L and diameter of 30 cm was anchored into the soil surface to collect gas emitted from the soil. This method detects the concentration change over time until the gas level in the chamber reaches a steady state. A DC pump (Motorbank, Korea) and an air flow meter (Dwyer, USA) were installed to maintain a constant flow rate ($\cong 1$ L min^{-1}) of air between the chamber and the detector, thereby forming a continuous air circulation system containing GHGs.

The concentration of CO_2 was directly determined using a CO_2 sensor (Soha-Tech, Korea) incorporated within the measurement unit. To measure N_2O, 30 mL of air was collected by syringe through the sampling port from a closed chamber. The sample was immediately transferred to pre-evacuated 12 mL vials (Labco, 839W, UK) for sample preservation during transport to the laboratory [24]. In the laboratory, the concentration of N_2O was determined using a gas chromatograph (Simadzu, GC-2010, Japan) equipped with an electron capture detector. N_2O peak separation was performed using a stainless-steel column packed with an 80/100 mesh Porapak Q (Agilent, CP-3800, USA). The carrier gas was N_2 at a flow rate of 30 mL min^{-1}, and the make-up gas was a mixture of 10% CO_2 in N_2 at 6 mL min^{-1}. The temperatures of the oven, injector, and detector were set to 55, 100, and 340 °C, respectively [4]. The volume of sample injection was 500 µL. Over 11 min running time, N_2O peak retention time was 5.8 min.

Field measurements were conducted twice on a given day (between 8:00–11:00 a.m. and between 3:00–6:00 p.m.) and the results were averaged to determine daily emissions based on the sampling protocols [24]. The flux of gas emissions (gas flux, F, mg m^{-2} hr^{-1}) was calculated by Equation (1) [4]:

$$F = \rho \cdot \left[\frac{V}{A}\right] \cdot \left[\frac{\Delta C}{\Delta t}\right] \cdot \left[\frac{273}{(T+273)}\right], \tag{1}$$

where ρ is the density of gas (mg m^{-3}), V is the volume of the chamber (m^3), A is the area of the bottom of the chamber (m^2), $\frac{\Delta C}{\Delta t}$ is the average rate of change in concentration (ppmV h^{-1}), and T is the average temperature in the chamber (°C).

2.3. Statistical Analysis

We employed the coefficient of determination (r^2) to validate the DNDC model performance, as shown in Equation (2) [16]:

$$r^2 = \left(\frac{\sum_{i=1}^{n}(M_i - \overline{M})(P_i - \overline{P})}{\sqrt{\sum_{i=1}^{n}(M_i - \overline{M})^2 \cdot \sum_{i=1}^{n}(P_i - \overline{P})^2}}\right)^2, \tag{2}$$

where P_i is the predicted value, M_i is the measured value, $i = 1, \ldots, n$ is the number of measured values, and $\overline{P}$ and $\overline{M}$ are the means of the predicted and measured values, respectively. Statistical differences were determined at the significance level of $p < 0.001$ for between field measurements and model prediction and at $p < 0.05$ for between farming variables using SAS 9.4 (SAS Institute Inc., USA).

3. Results and Discussion

3.1. Field Verification of DNDC Model Predictions

Air temperature and the magnitude of rainfall during the field study are shown in Figure 1a. The annual mean temperature was 12.6 ± 2.4 °C and the annual precipitation was 1467 ± 185 mm. Rainfall was concentrated between June and September, representing a typical East Asian monsoon climate. Heavy rainfall (e.g., >110 mm according the criteria of Hong [25]) was recorded twice on the weeks of June 26 and August 29 at 118 mm and 279 mm, respectively. The period of cabbage cultivation recommended by the KRDA is between August 11 and October 31. Figure 1b,c compares the model GHGs outputs with selected field measurements under the T1F3 farming practice. The annual flux of CO_2 and N_2O from the cabbage field were 19.3 ± 0.5 t CO_2-eq ha^{-1} yr^{-1} and 4.4 ± 0.6 t CO_2-eq ha^{-1} yr^{-1}, respectively, and the daily flux of GHGs peaked in the week of heavy rainfall. During the non-cropping season, N_2O flux was negligible while emissions of CO_2 continued.

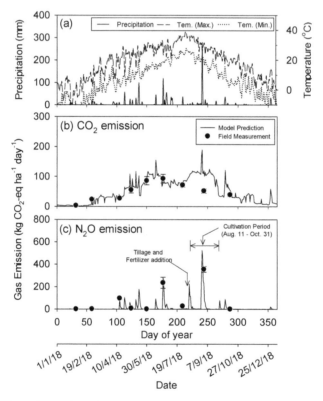

Figure 1. Climate data and greenhouse gas emissions from the Deokso cabbage field during the experimental period (2018). Daily maximum and minimum air temperatures and daily precipitation are shown in plot (**a**). Model simulations of CO_2 and N_2O emissions are shown in plots (**b**,**c**), respectively, along with the field measurement data. The timing of tillage and fertilizer addition (August 4) and the duration of cabbage farming (August 11 to October 31) are also indicated in plot (**c**).

Field measurements of daily CO_2 and N_2O emissions (n = 36 for each) under four farming practices (T1F1, T3F1, T1F3, and T3F3) are plotted with their respective model outputs in Figure 2. We determined r^2 values of 0.70 ($p < 0.001$) for CO_2 and 0.89 ($p < 0.001$) for N_2O through linear regression analysis of the two datasets. The coefficient of determination (r^2) signifies how accurately field measurements are replicated by the model prediction. The 1:1 line is represented as dotted lines. Our r^2 results therefore indicate that the DNDC model can accurately predict GHGs emissions from cabbage fields under the different farming practices investigated in this study.

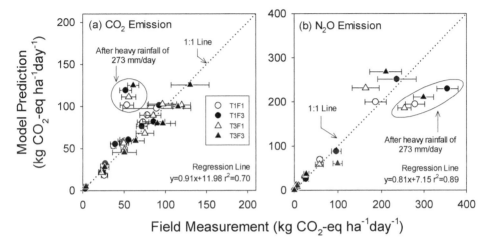

Figure 2. Comparisons of (**a**) CO_2 emissions and (**b**) N_2O emissions between the field-measurements from the Deokso cabbage field and the model predictions under four sets of farming practices (T1F1, T1F3, T3F1, and T3F3); T1 and T3 denote 10 cm and 30 cm tillage depth, respectively, and F1 and F3 denote 100 kg N ha^{-1} and 400 kg N ha^{-1}, respectively. Each data point represents the mean of triplicates measured in a given day. The error bars represent the standard deviations. The dotted line represents the 1:1 line.

However, we observed an apparent deviation from the correlation in the week of heavy rainfall (Figure 2). We observed slightly higher CO_2 flux estimates and lower N_2O flux estimates relative to the field measurements. This deviation may be attributed to the underestimation of soil volume wetness (%). The model prediction of soil wetness is well correlated with field measurements throughout the year (Figure S2). However, during the week of heavy rainfall (e.g., August 29), the modelled soil wetness (~22%) was significantly lower than the field measurements (~33%). The model exaggeration of soil drainage will overestimate the oxygen supply for microbial degradation of soil organic matter (SOM), which in turn will overestimate soil CO_2 production [26]. Conversely, the model could underestimate denitrification due to its overestimation of redox potential [11,18].

3.2. Modeling Results under Different Farming Practices

Model results for annual cabbage yield (t ha^{-1} yr^{-1}) and annual GHGs emissions (t CO_2-eq ha^{-1} yr^{-1}) under the RCP 8.5 scenario and as a function of the nine farming practices are shown in Table 2. The results of the KRDA's conventional farming method (i.e., T1F3; 10 cm tillage depth and 400 kg N ha^{-1} fertilization) are italicized. Further, we also illustrate the results of the model baseline, which assumes no climate change under conventional practices (T1F3).

Table 2. The model outputs for cabbage yield and greenhouse gas emissions from Korean cabbage fields under nine farming practices combining variations in tillage depth and levels of N-fertilizer. The model was simulated under the Representative Concentration Pathways 8.5 scenario for the 2020s and 2090s.

Model Outcome / Farming Practice [a]	Cabbage Yield (t ha^{-1} yr^{-1})		Greenhouse Gas Emission (t CO$_2$-eq ha^{-1} yr^{-1}) [d]			
			CO$_2$		N$_2$O	
	2020s	2090s	2020s	2090s	2020s	2090s
T1F1	34.8 ± 3.0 [b]	67.6 ± 5.1	9.3 ± 0.4	9.9 ± 0.4	2.8 ± 0.4	3.4 ± 0.3
T1F2	55.2 ± 3.7	94.7 ± 5.4	9.8 ± 0.5	10.2 ± 0.4	4.3 ± 0.5	5.1 ± 0.4
T1F3 *(Conventional farming practice)*	*65.4 ± 3.8*	*103.4 ± 6.7*	*10.1 ± 0.4*	*10.4 ± 0.4*	*6.2 ± 0.6*	*6.9 ± 0.5*
T2F1	38.7 ± 3.3	69.2 ± 5.1	10.0 ± 0.5	10.6 ± 0.5	2.4 ± 0.3	3.2 ± 0.3
T2F2	56.6 ± 3.8	94.5 ± 5.7	10.7 ± 0.5	11.0 ± 0.5	3.8 ± 0.4	4.8 ± 0.4
T2F3	64.2 ± 3.9	108.6 ± 5.7	10.9 ± 0.6	11.2 ± 0.5	5.8 ± 0.5	6.7 ± 0.5
T3F1	50.4 ± 3.3	78.1 ± 5.0	11.3 ± 0.6	11.6 ± 0.5	2.2 ± 0.3	3.3 ± 0.3
T3F2	59.7 ± 3.5	94.7 ± 5.1	11.7 ± 0.6	11.9 ± 0.5	3.7 ± 0.4	5.0 ± 0.4
T3F3	65.8 ± 3.5	104.3 ± 5.2	11.8 ± 0.6	12.1 ± 0.6	5.7 ± 0.5	7.0 ± 0.5
Baseline [c] (No climate change with T1F3)	63.0 ± 3.4	61.2 ± 5.1	9.8 ± 0.5	9.8 ± 0.5	6.1 ± 0.4	6.1 ± 0.5

a. The results for the farming practice recommended by the Korea Rural Development Administration (KRDA) for cabbage cultivation (T1F3) is italicized. b. Values following the ± sign denote standard deviation. c. Model results assuming no climate change under the KRDA farming method (T1F3). d. N$_2$O emissions were converted to the unit of CO$_2$-equivalents (CO$_2$-eq).

3.2.1. Impacts of Farming Practices on Cabbage Yield and GHGs Emissions

Projected cabbage yield (t ha^{-1} yr^{-1}) and GHGs emissions (t CO$_2$-eq ha^{-1} yr^{-1}) for the 2020s and 2090s varied widely depending on the farming practice applied. For a given fertilizer level, the predicted cabbage yield had increased with increasing tillage depth (T1 to T3). Likewise, for a given tillage depth, higher N-fertilizer application increased cabbage yield. Thus, we projected highest cabbage yields under the T3F3 farming practice for both the 2020s and 2090s as 65.8 t ha^{-1} yr^{-1} and 104.3 t ha^{-1} yr^{-1}, respectively. We found the impact of N-fertilizer levels on cabbage yield to be significantly different within the range of the two farming variables (significance level of $\alpha = 0.05$). For example, increasing fertilizer application (from F1 to F3) resulted in an 87.9% (30.6 t ha^{-1} yr^{-1}) increase in cabbage yield for T1 in the 2020s. In contrast, increasing tillage depth (from T1 to T3) resulted in only a 44.8% increase in yield (15.6 t ha^{-1} yr^{-1}) for the lowest fertilizer level (F1) in the same period. Further, tillage depth became less impactful on yield at conventional levels of N-fertilizer input (F3). For example, the projected cabbage yields for T1F3, T2F3, and T3F3 showed no statistical difference ($\alpha = 0.05$) in both the 2020s and 2090s. Our results suggest that the impacts of deep tillage on cabbage production are insignificant when adequate levels of N are supplied.

Both CO$_2$ and N$_2$O emissions increased concurrently with increasing tillage depth and N-fertilizer levels (Table 2). The flux of CO$_2$ was therefore highest under T3F3. Within the range of variables in this study, tillage depth had a larger influence on CO$_2$ emissions. For example, for a given tillage depth, increasing fertilizer addition (from F1 to F3) led to a CO$_2$ increase of ≤0.9 t CO$_2$-eq ha^{-1} yr^{-1}, while increasing tillage depth from T1F3 to T3F3 led to an increase of 1.7 t CO$_2$-eq ha^{-1} yr^{-1} (~17%) in the 2020s. Similar CO$_2$ emissions patterns were observed for the 2090s. However, cabbage yield remained statistically constant with increasing tillage depth under conventional levels of N-fertilizer application. Deeper tillage can promote CO$_2$ production, as it improves air supply to heterotrophic microbes for SOM degradation [10,12].

In contrast to CO_2, the flux of N_2O was highest under T1F3. N_2O emissions had increased with increasing N-fertilizer addition but decreased with increasing tillage depth. For example, in the 2020s, the level of N_2O emissions occurred in the following order: T1F3 > T1F2 > T1F1 and T1F3 > T2F3 > T3F3. Nitrogen-fertilizers are a main source of N_2O formation during the denitrification process in agricultural ecosystems [11]. In the DNDC model, urea ($CO(NH_2)_2$) application undergoes a series of biochemical reactions in soil as follows: urea $\rightarrow NH_3 \rightarrow NO_3^- \rightarrow NO_2^- \rightarrow NO \rightarrow N_2O$. Each nitrogen-loss process is performed primarily by heterotrophic bacteria in anoxic environments. Thus, deep tillage practices can limit the formation of N_2O due to rapid gas exchange between the soil and atmosphere.

3.2.2. Impacts of Climate Change on Cabbage Yield and GHGs Emissions

Under the RCP 8.5 climate change scenario, both cabbage yield and GHGs emissions are significantly higher in the 2090s relative to the 2020s regardless of the farming practice applied. The model projects a 28–44 t ha^{-1} yr^{-1} increase in cabbage yield from the 2020s to the 2090s depending on the farming practice applied. We observe increases of 0.2–0.6 t CO_2-eq ha^{-1} yr^{-1} for CO_2 emissions and 0.6–1.3 t CO_2-eq ha^{-1} yr^{-1} for N_2O emissions (Table 2). Further, we observed no significant difference in the outputs of the model baseline between the two time periods ($\alpha = 0.05$).

According to the RCP 8.5 scenario, the model predicts the mean temperature and precipitation of the 2090s cropping season to be 23.2 °C and 281 mm, respectively, which are significantly higher than the baseline values at 18.0 °C and 242 mm, respectively. If sufficient nutrients are supplied, the physiological activity of cabbage plants are enhanced under these predicted climatic conditions, leading to increased farmland productivity [14,17]. Note that the annual cabbage yield increases with increased fertilization in the order of F3 ≥ F2 >> F1. Under conventional farming methods, cabbage yield is expected to increase by 38 t ha^{-1} yr^{-1} from the 2020s to the 2090s.

Within the range of variables considered in this study, the percentage rise in N_2O is higher than the rise in CO_2 between the two time periods. N_2O emissions in the 2090s were 11–50% higher than the 2020s, while the rise in CO_2 emissions were ≤6% higher than the 2020s. The production of CO_2 and N_2O from the microbial decomposition of SOM and the denitrification of N-fertilizers, respectively, are both temperature dependent processes [14]. The observed temperature increase following the RCP 8.5 projections suggests that the activity of microbial denitrification is expected to increase by 2–4 times, while the activity of microbial SOM decomposition will be less affected [27]. We would therefore expect a greater rise in N_2O relative to CO_2 emissions due to the different temperature dependencies of the biochemical reactions [14,18]. Previous modeling studies have also reported similar future GHGs emissions predictions of upland fields under various climate scenarios [6,7,13]. For example, CO_2 was found to be the dominant cause of future temperature rise and was projected to increase by 5%–14% from the 1990s to the 2090s [6]. In agreement, another study projected an increase in Canadian N_2O emissions of 9.5%–31.2% in 2071–2100 relative to baseline levels in 1971–2000 [13].

3.3. Model Results of the Best Farming Practices

The model outputs for cabbage yield and GHGs emissions optimized to best achieve the three scenario goals (i.e., minimizing GHGs, maximizing yield, and maintaining demand) in the 2020s and 2090s are presented in Table 3. For comparison, the model outputs from the conventional method (i.e., T1F3) are also shown in Table 3. Under the conventional method, cabbage yield in the 2020s is almost equal to the demand, while a cabbage yield of 33.9 t ha^{-1} yr^{-1} exceeds demand in the 2090s. Table S1 shows the optimum distribution of the nine farming practices over the 8072 cabbage fields to achieve each scenario goal.

Table 3. Model outputs for cabbage yield and greenhouse gas (GHG) emissions when each of the three scenario goals (minimizing GHGs, maximizing yield, and maintaining demand) are achieved under the Representative Concentration Pathways 8.5 projection.

Model Outcome Scenario Goals [a]	Cabbage Yield (t ha^{-1} yr^{-1})		GHGs Emission (t CO$_2$-eq ha^{-1} yr^{-1}) [c]	
	2020s Demand Forecasting = 65.1 ± 3.3	2090s Demand Forecasting = 74.5 ± 3.7	2020s	2090s
Minimizing GHGs	35.5 ± 0.3 [b]	68.6 ± 0.2	12.0 ± 0.1 (−26.4%) [d]	13.3 ± 0.1 (−23.1%)
Maximizing Yield	68.1 ± 1.3	109.2 ± 1.5	17.3 ± 0.4 (+6.13%)	18.6 ± 0.3 (+7.51%)
Maintaining Demand	65.2 ± 1.6	74.8 ± 2.0	16.0 ± 0.4 (−1.84%)	13.9 ± 0.3 (−19.6%)
Conventional Method	64.5 ± 3.8	103.4 ± 6.7	16.3 ± 1.1	17.3 ± 0.9

a. The scenario goals were achieved by allocating one of the nine farming practices into the 8072 Korean cabbage field cells for the particular time period. Minimizing GHGs = farming practices to achieve the minimum GHG emission. Maximizing yield = farming practices to achieve the maximum cabbage yield. Maintaining demand = farming practices to balance future cabbage yield with future demand for the particular time period. Conventional method = farming practices which meet the requirements of the KRDA (10 cm tillage and 400 kg N ha^{-1} fertilizer). b. Values following the ± sign denote the standard deviations. c. Sum of CO$_2$ and N$_2$O emissions given as units of CO$_2$-equivalents. d. Value in parentheses denotes the percentage reduction (−) or percentage increase (+) in GHGs emissions relative to the conventional method.

The minimum cabbage field GHGs emission was achieved by adopting low-carbon farming practices (e.g., 75% of T1F1, 22% of T2F1, and 3% of T3F1 for the 2020s and 94% of T1F1 and 6% of T2F1 for the 2090s; Table S1). This particular distribution of farming practice would reduce GHGs emissions by 26.4% and 23.1% in the 2020s and 2090s, respectively, compared to conventional farming. However, this scenario is unrealistic, as cabbage production is projected to be lower than the forecast for demand. A smaller difference between yield and demand was observed in the 2090s relative to the 2020s, as climate change favors cabbage growth.

We achieved maximum cabbage yield through deep tillage and high fertilizer levels (e.g., 7% of T2F3 and 64% of T3F3 for the 2020s, and 52% of T2F3 and 45% of T3F3 for the 2090s; Table S1). Under this scenario, cabbage yield had exceeded demand. GHGs emissions had increased by 6.13% and 7.51% in the 2020s and 2090s, respectively, due to enhanced CO$_2$ production caused by deep tillage practices (T2 and T3).

The balance between demand and supply must be considered in order to maintain future cabbage demand. Cabbage yield under the conventional method was predicted to be almost equal to the forecast for demand in the 2020s (64.5 t ha^{-1} yr^{-1} vs. 65.1 t ha^{-1} yr^{-1}; Table 3). We achieved this goal by selectively adopting the high-yield and low-carbon farming practice on 48% of cabbage fields (Table S1). Therefore, half of the total cabbage fields retained the conventional farming practice, and its total contribution towards GHGs mitigation was negligible (1.84%). The cabbage production under conventional farming (103.4 t ha^{-1} yr^{-1}) was greater than the forecast for demand (74.5 t ha^{-1} yr^{-1}) in the 2090s. To balance supply and demand, we therefore adopted low fertilizer farming practices (e.g., 57% of T1F1, 6% of T2F1, 17% of T3F1, etc.; Table S1) on 94% of the cabbage fields, which resulted in a 19.6% reduction in GHGs emissions (Table 3). However, it should be noted that farmers who adopt these low-carbon farming practices (approximately 7112 fields) in place of the conventional method are likely to experience profit-loss due to a reduction in cabbage yield.

The GHGs distribution of the 8072 cabbage fields in the 2090s for each scenario are displayed in Figure 3. The GHG distribution under the conventional practice is also shown as dashed bars for comparison. The data are fairly symmetric about the mean for all cases, indicating that the emissions and yield data are normally distributed. The conventional farming dataset (plot (a)) has the highest standard deviation and therefore widest spread relative to the other scenarios. We achieved both minimum GHGs emissions (plot (b)) and a balanced cabbage demand and supply (plot (d)) by converting cabbage fields with high GHGs emissions (mostly high yields) to low emission fields (mostly low yields). We achieved maximum yield by converting low yield cabbage fields (but potentially productive by changing farming methods) to high yield fields.

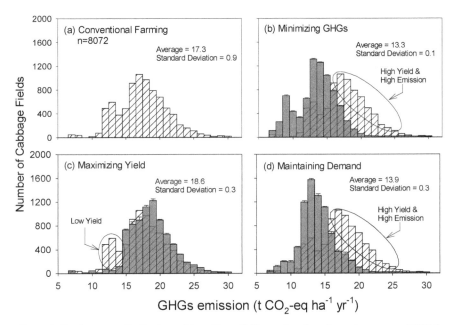

Figure 3. The frequency distributions of the cabbage fields as a function of greenhouse gases (GHGs) emissions to achieve the three scenario goals under the Representative Concentration Pathways 8.5 scenario for the 2090s, including distributions for (**a**) conventional farming, (**b**) minimizing GHGs emissions, (**c**) maximizing cabbage yield, and (**d**) balancing future cabbage demand and supply. The results of the conventional farming practice (dashed white bars) are also displayed in plots (**b–d**) for comparison.

3.4. Implications

Our results demonstrate that current conventional farming practices (e.g., 400 kg N ha^{-1} fertilization and 10 cm tillage depth) under the RCP 8.5 scenario will produce cabbage yields of 103.4 t ha^{-1} yr^{-1} in the 2090s, which is 38.8% greater than the forecast for demand (74.5 t ha^{-1} yr^{-1}). The simulation results suggest that future cabbage demand can be met even when 90% of all cabbage fields adopt low-carbon farming practices. However, a disadvantage is that low-carbon farming practices may lead to a loss in profit due to a decrease in crop yield [28].

The Korean government compensates farmers who conduct low-carbon agriculture through the implementation of the Direct Payment Program for Low-Carbon Farming Practices [20]. This policy assesses both the potential benefits and income reduction as a result of choosing low-carbon farming practices over conventional methods, so that farmers are effectively compensated for their contributions towards the mitigation of GHGs emissions. The results of this study are therefore useful for the Korean government to quantify both the impacts of low-carbon practices on GHGs reduction and to mitigate against potential financial losses in the agricultural sector.

4. Conclusions

Results of this study demonstrate that the adoption of low-carbon farming practices (optimal N-fertilizer levels and tillage depth) can effectively reduce national GHGs emissions from cabbage fields without compromising future demand. The economic benefits to farmers will likely be compromised, but the government has incentives in place to account for financial losses. In Korea, farmers are incentivized to adopt low-carbon farming practices through the compensation of any resulting income loss via the government's direct payment program. In practice, the result of this study will aid the government to (i) effectively evaluate the contribution of low-carbon practices on GHGs mitigation from cabbage fields, and (ii) quantify yield and farmer profit losses in response to the adoption of low-carbon farming practices for accurate subsidy assessments and calculations. The conceptual framework of this modeling approach can be widely adopted and applied to other similar upland cropping systems. Our results demonstrate that optimizing both fertilization and tillage depth are effective strategies towards achieving the economical and sustainable management of Korean cabbage fields in response to future climate change.

Supplementary Materials: The following are available online at http://www.mdpi.com/2071-1050/11/21/6158/s1, Table S1: Optimum distributions of the nine farming practices over the 8072 cabbage fields to achieve the three scenario goals under Representative Concentration Pathways 8.5 for the 2020s and 2090s, Figure S1: (a) The geographical location of South Korea and the distribution of cabbage cultivation fields across the country; and (b) the illustrative components of the CO_2 and N_2O flux measurement system (not to scale), Figure S2: Comparisons between the Deokso cabbage field measurements and model predictions of soil volume wetness (%) under four sets of farming practice (T1F1, T1F3, T3F1, and T3F3). The data points represent the mean of triplicates measured in a given day. The error bars represent the standard deviations. The dotted line represents the 1:1 line. Acronyms can be referred to in the Figure 2 caption.

Author Contributions: Conceptualization, J.-G.K.; Data curation, W.H. and M.P.; Formal analysis, W.H. and S.H.; Investigation, M.P.; Methodology, M.P. and K.C.; Supervision, S.H.; Writing—original draft, W.H.; Writing—review & editing, S.H.

Funding: This study was in part funded by the Korea Ministry of Environment (MOE) as "Climate Change Correspondence Program (Project No. 2014-001310008)".

Acknowledgments: This study was in part funded by Korea University Grant.

Conflicts of Interest: The authors declare no conflict of interest.

References

1. Intergovernmental Panel on Climate Change (IPCC). *Climate Change 2013: The Physical Science Basis*; Contribution of Working Group I to the Fifth Assessment Report of the IPCC: Cambridge, UK; New York, NY, USA, 2013; p. 493.
2. Food and Agriculture Organization (FAO). *FAOSTAT Database*; FAO: Rome, Italy, 2012; Available online: http://fao.org/faostat/en/#home (accessed on 22 August 2019).
3. Intergovernmental Panel on Climate Change (IPCC). *Climate Change 2007: Synthesis Report*; Contribution of Working Groups I to the Fourth Assessment Report of the IPCC: Geneva, Switzerland, 2007.
4. Li, H.; Qiu, J.; Wang, L.; Tang, H.; Li, C.; Van Ranst, E. Modelling impacts of alternative farming management practices on greenhouse gas emissions from a winter wheat–maize rotation system in China. *Agric. Ecosyst. Environ.* **2010**, *135*, 24–33. [CrossRef]
5. Greenhouse Gas Inventory and Research Center (GIR). *National Greenhouse Gas Inventory Report of Korea*; GIR: Seoul, Korea, 2016. (In Korean)
6. Abdalla, M.; Kumar, S.; Jones, M.; Burke, J.; Williams, M. Testing DNDC model for simulating soil respiration and assessing the effects of climate change on the CO_2 gas flux from Irish agriculture. *Glob. Planet. Chang.* **2011**, *78*, 106–115. [CrossRef]
7. Smith, W.N.; Grant, B.B.; Desjardins, R.L.; Kroebel, R.; Li, C.; Qian, B.; Worth, D.E.; McConkey, B.G.; Drury, C.F. Assessing the effects of climate change on crop production and GHG emissions in Canada. *Agric. Ecosyst. Environ.* **2013**, *179*, 139–150. [CrossRef]
8. Asano, K.; Yang, H.; Lee, Y.; Yoon, J. Designing optimized food intake patterns for Korean adults using linear programming (I): Analysis of data from the 2010–2014 Korea National Health and Nutrition Examination Survey. *J. Nutr. Health* **2018**, *51*, 73–86. [CrossRef]
9. Korean Statistical Information (KOSIS). *Service Cultivated Area of Food Crops*; KOSIS: Daejeon, Korea, 2015. (In Korean)

10. Forte, A.; Fiorentino, N.; Fagnano, M.; Fierro, A. Mitigation impact of minimum tillage on CO_2 and N_2O emissions from a Mediterranean maize cropped soil under low-water input management. *Soil Tillage Res.* **2017**, *166*, 167–178. [CrossRef]

11. Snyder, C.S.; Bruulsema, T.W.; Jensen, T.L.; Fixen, P.E. Review of greenhouse gas emissions from crop production systems and fertilizer management effects. *Agric. Ecosyst. Environ.* **2009**, *133*, 247–266. [CrossRef]

12. Khalil, M.I.; Rahman, M.S.; Schmidhalter, U.; Olfs, H.W. Nitrogen fertilizer-induced mineralization of soil organic C and N in six contrasting soils of Bangladesh. *J. Plant. Nutr. Soil Sci.* **2007**, *170*, 210–218. [CrossRef]

13. He, W.; Yang, J.Y.; Drury, C.F.; Smith, W.N.; Grant, B.B.; He, P.; Qian, B.; Zhou, W.; Hoogenboom, G. Estimating the impacts of climate change on crop yields and N_2O emissions for conventional and no-tillage in Southwestern Ontario, Canada. *Agric. Syst.* **2018**, *159*, 187–198. [CrossRef]

14. Li, C.; Frolking, S.; Frolking, T.A. A model of nitrous oxide evolution from soil driven by rainfall events: 1. model structure and sensitivity. *J. Geophys. Res.* **1992**, *97*, 9759–9776. [CrossRef]

15. Li, H.; Qiu, J.; Wang, L.; Yang, L. Advance in a terrestrial biogeochemical model-DNDC model. *Acta Ecol. Sin.* **2011**, *31*, 91–96. [CrossRef]

16. Yun, J.I. Predicting regional rice production in South Korea using spatial data and crop-growth modeling. *Agric. Syst.* **2003**, *77*, 23–38. [CrossRef]

17. Zhang, Y.; Li, C.; Zhou, X.; Moore, B. A simulation model linking crop growth and soil biogeochemistry for sustainable agriculture. *Ecol. Modell.* **2002**, *151*, 75–108. [CrossRef]

18. Oertel, C.; Matschullat, J.; Zurba, K.; Zimmermann, F.; Erasmi, S. Greenhouse gas emissions from soils—A review. *Chem. Erde-Geochem.* **2016**, *76*, 327–352. [CrossRef]

19. Zhuang, M.; Zhang, J.; Lam, S.K.; Li, H.; Wang, L. Management practices to improve economic benefit and decrease greenhouse gas intensity in a green onion-winter wheat relay intercropping system in the North China Plain. *J. Clean. Prod.* **2019**, *208*, 709–715. [CrossRef]

20. Korea Rural Economic Institute (KREI). *Introducing Direct Payment Program for Low-Carbon Farming Practices and Creating an Action Plan. Guideline*; KREI: Naju, Korea, 2013. (In Korean)

21. Hewitt, H.T.; Copsey, D.; Culverwell, I.D.; Harris, C.M.; Hill, R.S.R.; Keen, A.B.; McLaren, A.J.; Hunke, E.C. Design and implementation of the infrastructure of HadGEM3: The next-generation Met Office climate modelling system. *Geosci. Model. Dev.* **2011**, *4*, 223–253. [CrossRef]

22. Korea Rural Economic Institute (KREI). *Vision of 2030/2050 Agriculture and Rural Sector in Korea*; KREI: Naju, Korea, 2010. (In Korean)

23. Hwang, W.; Kim, C.; Cho, K.; Hyun, S. Characteristics of Emission of Greenhouse Gases (CO_2 and CH_4) from Rice Paddy Fields in South Korea under Climate Change Scenario (RCP-8.5) using the DNDC Model. *Pedosphere* **2019**, in press.

24. Parkin, T.B.; Venterea, R.T. Chapter 3. Chamber-based trace gas flux measurements. In *GRACEnet Sampling Protocols*; Follett, R.F., Ed.; U.S. Department of Agriculture: Washington, DC, USA, 2010; pp. 1–39.

25. Hong, S.G. A study on the threshold values of heavy rain warning in Korea. *Asia-Pac. J. Atmos. Sci.* **1999**, *35*, 178–192.

26. Ghezzehei, T.A.; Sulman, B.; Arnold, C.L.; Bogie, N.A.; Berhe, A.A. On the role of soil water retention characteristic on aerobic microbial respiration. *Biogeosciences* **2019**, *16*, 1187–1209. [CrossRef]

27. Castaldi, S. Responses of nitrous oxide, dinitrogen and carbon dioxide production and oxygen consumption to temperature in forest and agricultural light-textured soils determined by model experiment. *Biol. Fertil. Soils* **2000**, *32*, 67–72. [CrossRef]

28. Kragt, M.E.; Gibson, F.L.; Maseyk, F.; Wilson, K.A. Public willingness to pay for carbon farming and its co-benefits. *Ecol. Econ.* **2016**, *126*, 125–131. [CrossRef]

Article

Sustainable Production of Sweet Sorghum as a Bioenergy Crop Using Biosolids Taking into Account Greenhouse Gas Emissions

Lilianna Głąb * and Józef Sowiński

Institute of Agroecology and Plant Production, Wroclaw University of Environmental and Life Sciences 24a Grunwaldzki Square, 50-363 Wrocław, Poland; jozef.sowinski@upwr.edu.pl
* Correspondence: lilianna.glab@upwr.edu.pl

Received: 14 April 2019; Accepted: 24 May 2019; Published: 29 May 2019

Abstract: Currently, little data are available on greenhouse gas (GHG) emissions from sweet sorghum production under temperate climate. Similarly, information on the effect of bio-based waste products use on the carbon (C) footprint of sorghum cultivation is rare in the literature. The aim of this study was to evaluate the agronomical and environmental effects of the application of biosolids as a nitrogen source in the production of sweet sorghum as a bioenergy crop. The yield of sorghum biomass was assessed and the GHG emissions arising from crop production were quantified. The present study focused on whether agricultural use of sewage sludge and digestate could be considered an option to improve the C footprint of sorghum production. Biosolids—sewage sludge and digestate—could be recognized as a nutrient substitute without crop yield losses. Nitrogen application had the greatest impact on the external GHG emissions and it was responsible for 54% of these emissions. CO_{2eq} emissions decreased by 14 and 11%, respectively, when sewage sludge and digestate were applied. This fertilization practice represents a promising strategy for low C agriculture and could be recommended to provide sustainable sorghum production as a bioenergy crop to mitigate GHG emissions.

Keywords: GHG emissions; carbon footprint; sweet sorghum; fertilization management; digestate; sewage sludge

---

1. Introduction

Sustainable agricultural systems should be economically profitable, but they should also provide food, feed, and biofuels and prevent or even enhance ecosystem services [1]. In recent years, the agricultural sector has become increasingly heavily dependent on mainly non-renewable energy sources. One of the challenges for sustainable crop production is to decrease the external energy inputs [2]. Among the key issues for agricultural sustainability is greenhouse gas (GHG) emissions as well as their effect on climate [3]. Agricultural production has a significant impact on climate change [4]. Emissions from the agricultural sector in the European Union were estimated at 432 million tons of CO_2 equivalents (CO_{2eq}) in 2017 and this was responsible for emitting 10% of the total amount of European GHG emissions [5]. Effective methods (i.e., methods which have the potential to mitigate emissions in agriculture) need development, because current decreases in these areas of emission levels are insufficient [6]. However, development of the following methods for carbon (C) reduction in agriculture are promising: precision farming; improved fertilization management; cultivation of crops with a higher potential for C sequestration (i.e., C4 photosynthesis cycle crops); and lastly, but not all-inclusively, the implementation of organic fertilizers and alternative soil amendments to replace synthetic fertilizers [7,8].

Varied management practices have different impacts on GHG emissions and crops cultivation and therefore they should be examined [9]. The detailed estimation of GHG emissions from the agricultural sector allows for the identification of hot-spots, which provide information about which input causes the most significant effect on climate change due to the release of GHG [10]. The amounts of carbon dioxide (CO_2), methane (CH_4), and nitrous oxide (N_2O) emissions from various sources are converted to one unit, such as kilograms of $CO_{2\,eq}$, emitted to the atmosphere and this is defined as the C footprint [11]. The C footprint can be quantified on a land-area basis as a spatial C footprint, on an output basis defined as per yield unit of produced biomass C footprint or on a produced energy basis [12,13].

In the European Union in 2009, legislation in the form of Renewable Energy Directives provided the criteria for biofuels [14,15]. These two legislative acts state that dedicated energy crops cultivation is one of the three main stages of biofuels production which should be taken into account during GHG emission quantification [16]. Carbon footprint calculators, which take into account the LCA approach represent useful tools for estimating GHG emissions from cultivation of crops for energy purposes. A detailed overview of a wide range of these calculators was provided by Peter et al. [10]. One such calculator—BioGrace (Biofuel Greenhouse Gas Emissions in Europe)—was used in the present study to quantify GHG emissions from energy crops such as sweet sorghum (*Sorghum bicolor* (L.) Moench) at farm stage [17]. This calculation tool is approved and recommended by the two key European Directives as a method to harmonize calculations of biofuel GHG emissions and support implementation of European directives [14,15] into national laws [16]. BioGrace was developed by economic operators for consultants and politicians to improve decision-making and for implementation of national programs aimed at decreasing GHG emissions [10,16]. It is also a useful tool for farmers, who can check how different management practices affect the carbon footprint of biofuel production [10]. The method of GHG estimation offered by the BioGrace tool is based on standard conversion values, which are mainly emission factors, and also other data that are needed to convert some agricultural inputs into emissions [16]. The Tier 1 approach with national or global standard values was used in this study. The country specific emission values allow improvement of these calculations according to the Tier 1 approach [18].

Greenhouse gas emissions can be divided into external and on-farm emissions [11]. These emissions are a result of production processes and application of agricultural inputs, such as pesticides, fertilizers, seeds, and combustion of diesel oil during farm operations [8,19]. Production and application of fertilizers is a significant contributor to the emissions of GHG from arable crop production [11,12,20,21]. Crop production should take into account the C footprint of the whole biomass energy production chain, in particular at the farm stage. A more sustainable approach to nitrogen (N) fertilization management has a large potential to decrease GHG emissions from energy crop cultivation [10]. There is a great need to focus on more sustainable improvement of soil fertility and optimal use of synthetic N fertilizers. Application of bio-based by-products represents a sustainable waste management method and it provides recycling nutrients for crop growth, which is in line with the European policy for a circular economy [2,22–24]. Results of studies concerning the environmental impact of digestate vary significantly throughout the literature. Agricultural use of digestate, which is a stable organic waste material, has the potential to decrease soil CO_{2eq} emissions [25]. However, there is an environmental risk associated with increasing N_2O volatilization [26]. Some authors have noted especially high N_2O emissions, when liquid digestate is applied in moderately wet soil [27,28]. This significant loss of N_2O can negate any benefits from replacing synthetic fertilizers with digestate [28]. Pezzolla et al. [29] obtained different results and suggest that the digestate can be used as a fertilizer to grow crops without any harmful effect on the agroecosystem, including GHG emissions. However, data about the impact of sewage sludge application as a soil amendment on the C footprint of crop cultivation is still lacking.

Biomass use for production of heat, electricity, and fuel has significantly increased in recent years [30]. Using energy crops with a high C sequestration potential such as sweet sorghum for energy purposes represents an almost-closed CO_2 cycle [8,10]. It is believed that the biofuels production processes give the same amount of CO_{2eq} as was fixed in the biomass through photosynthesis at the production stage of the raw material [31]. However, crop growth generates a certain amount of GHG and therefore it is not the ideal carbon neutral production chain [10,31]. Sweet sorghum represents a promising bioenergy feedstock in the temperate climate of Central Europe [32]. It is still not well understood how energy crop cultivation systems affect GHG emissions. Moreover, so far, there are a limited number of studies on GHG emissions from sorghum production. Storlien et al. [9] examined the effect of various N fertilizer rates, crop rotation, and crop residue managements on GHG emissions from sorghum production for bioenergy purposes. According to their results, N addition significantly increased N_2O emissions, and incorporation of half of sorghum residues increased CO_2 emissions [9]. Davis et al. [33] recommended the perennial grasses switchgrass (*Panicum virgatum* L.) and miscanthus (*Miscanthus* × *giganteus* Greef et Deuter) for ethanol instead of maize (*Zea mays* L.) as a cellulosic feedstock for ethanol production. The production of biofuel from these crops characterized by low N demand allows significant mitigation of GHG emissions at the farm stage [33].

Given the aforementioned considerations, the response of sweet sorghum dry matter (DM) yield to biosolids fertilization was evaluated. This study focused on evaluating how sewage sludge and digestate application affect the C footprint of the sorghum production system, compared with conventional fertilization management. This study improves knowledge on the environmental impact of sorghum production with different fertilization managements in regards to the C footprint. It provides insight into the necessity of finding the most sustainable and cleanest methods of crop production for energy purposes.

2. Materials and Methods

2.1. Study Site Description

This study on N fertilization management of sorghum yield and GHG emissions was carried out in the period 2016–2018. It was conducted under field experiment conditions at the Agricultural Research Station of Wroclaw University of Environmental and Life Sciences. This experimental site is located in Central Europe (southwestern Poland; 51°10′25″ N and 17°07′02″ E). The climate in the study site is classified as temperate [34]. The annual average temperature is 9.0 °C and the annual average rainfall is 583 mm. The monthly temperature and precipitation of the area of research at the time the experiment was carried out were recorded every 10 min using an 'AsterMet' meteorological station (A-STER s.c., Krakow, Poland).

Ten soil samples (0.1 kg) from a layer of 30 cm depth were randomly taken from 10 spots across the experimental area (PN-ISO 10381-2). The soil samples were collected before sowing, prior to the field experiment was established. Then, individual samples were mixed to receive one composite sample, air dried and sieved with a mesh size of 2.0 mm. Then, physio-chemical properties of the topsoil (layer of 0–30 cm depth) were assessed. Particle-size distribution was determined by the sieve method (sand fraction) and hydrometer method (silt and clay fractions) after sample dispersion with hexametaphosphate. The soil texture was defined as sand, 0.05–2.0 mm; silt, 0.002–0.05 mm, and clay, <0.002 mm diameter (PN-R-04033 I USDA). pH was measured in distilled water and 1 mol dm^{-3} KCl solution, at the soil:solution ratio 1:2.5 with a pH meter (Omega Engineering, Inc., Norwalk, USA).

Mineral forms of nitrogen (soluble and exchangeable NH_4^+ and NO_3^-) were determined after extraction of fresh soil (stored in a fridge until extraction at −20 °C temperature) with a 1% solution of K_2SO_4, at the solution to soil ratio of 5:1 and a shaking time of 1 h. Then, suspensions of each derived sample were prepared by filtration through Whatman 2 filter paper (Whatman International Ltd, Maidstone, UK), followed by storage at −20 °C until analysis. In the filtrate, concentrations of mineral nitrogen forms were measured colorimetrically: N-NO_3 with phenoldisulfonic acid, and N-NH_4 with potassium sodium tartrate and Nessler's reagent (UV-Vis spectrometer, Cintra 4040, GBC Scientific Equipment, Braeside, Australia) [35]. The contents of plant-available phosphorus and potassium were analyzed by ICP-OES after Egner–Riehm extraction with calcium lactate (spectrometer Varian Inc. (Part A)—Vista MPX Simultaneous ICP-OES) [35].

2.2. Experimental Materials

Sucrosorgo 506—a late-maturing photoperiod insensitive triple-cross hybrid of sweet sorghum developed by Sorghum Partners Inc. (USA)—was used in this experiment. Stems of this hybrid have a relatively higher concentration of soluble sugars. It is well adapted for Central European conditions and produces a high amount of biomass in moderate climates [36]. Medium maturing, triple-cross hybrid Rona 1 with juicy stems recommended for silage is the result of a Hungarian breeding program (Gabonakutató) [37]. French late-maturing hybrid SuperSile 20 was received from Caussade Semences [38]. Late-maturing Goliath, mainly used for forage, was developed by Saaten Union GmbH (Germany) [39]. These hybrids have been registered in the EU Common Catalogue and were chosen based on the results of previous long-term studies [36].

Commercial urea characterized by 46% N was used in this study. Triple superphosphate with 46% P_2O_5 content was used as a phosphorus (P) source and potassium salt with 60% K_2O content as a potassium (K) fertilizer. Solar dried sewage sludge sourced from a municipal sewage treatment plant in Klodzko operating in EXPOVAL technology was used. Methanogenic post-digestion liquid digestate (termed digestate in this paper) was obtained from a mesophilic biogas plant in Strzelin (Südzucker Polska Inc.) fed with beet pulp. The chemical characteristics of biosolids used for fertilization are given in Table 1.

Table 1. Characteristics of biosolids used as fertilizers in the field experiment.

Parameters/Chemical Elements with Limit Value for Organic Fertilizer and Organic Soil Improver	Unit	Digestate	Sewage Sludge	Methods
pH		7.6	7.4	PN-EN 12176:2004
DM [1]	%	2.8	42	PN-EN 12880:2004
Organic compounds		71	31.5	PN-EN 12879:2004
Total nitrogen (N)		8	1.29	KJ-I-5.4-179
Ammonia nitrogen N-NH4	% DM	2	<0.10	PN-EN 14671:2007
Total phosphorus (P)		0.54	1.63	
Calcium (Ca)		2.99	4.11	PN-EN ISO 1185:2009
Magnesium (Mg)		1.02	0.60	
Potassium (K)		1280	n.a. [2]	PN-EN ISO 1185:2009; PN-EN 13657:2006
Copper (Cu), 200 [3]		49.6	239	
Zinc (Zn)		170	777	
Lead (Pb), 120 [3]	mg kg^{-1} DM	6.13	94	PN-EN ISO 1185:2009
Cadmium (Cd), 3 [3]		2.78	0.71	
Chromium (Cr)		11.2	32.9	
Nickel (Ni), 50 [3]		11.6	24.7	
Mercury (Hg), 1 [3]		0.050	0.540	KJ-I-5.4-36
Salmonella bacteria:				
no *Salmonella* species in 25 g sample [3]		0	0	PB/BB/7/F:20.03.2014

The results received from Südzucker Polska S.A. and Wodociagi Klodzkie sp. z o.o.; [1] DM dry matter; [2] n.a. not analyzed; and [3] the maximum permissible concentrations of contaminants in organic soil improver in the framework of the Fertilizing Product Regulation Proposal for a Regulation on the making available on the market of CE marked fertilizing products and amending Regulations (EC) No 1069/2009 and (EC) No 1107/2009 [22].

2.3. Field Treatments and Experimental Design

The experiment had a two-factorial split-plot design including sorghum hybrids and four fertilization managements. Treatments were arranged in a randomized, complete block with four replications. Experimental plots were 12.6 m^2 6 × 2.1 m (length × width). The research used four sweet sorghum hybrids: Sucrosorgo 506, Rona 1, SuperSile 20, and Goliath.

Primary tillage was done with a moldboard plow in the fall and with leveling and an aggregate seedbed preparation in the spring. The seeds of sweet sorghum hybrids were sown on 6 May 2016, 5 May 2017, and 15 May 2018. The annual N input of 100 kg N ha^{-1} was provided by broadcast application before sowing of (1) 19 t ha^{-1} sewage sludge, (2) 45 m^3 ha^{-1} digestate, and (3) 220 kg ha^{-1} urea. Unfertilized plots were also included in the experimental work. The rates of potassium and phosphorus were as follows: 100 kg ha^{-1} in the form of K$_2$O and 70 kg ha^{-1} of P$_2$O$_5$. These fertilizers were provided by single broadcast pre-sowing application and then were mixed with topsoil using a rotary harrow. Additional surrounding plots were set to minimize boundary effects through edges of the experimental field. Lumax 537.5 SE (s-metolachlor 312.5 g·dm^{-3} + mesotrione 37.5 g·dm^{-3} + terbuthylazine 187.5 g·dm^{-3}) at the dose of 2.0 dm^3·ha^{-1} was post-sowing, pre-emergency applied for weed control. Safener Concep III (oxabetrinil 700 g kg^{-1}) was used to avoid injuries to sorghum plants caused by the phytotoxic effects of s-metolachlor.

Before harvest, five representative plants from the middle row within each plot were collected and the aggregate sample was cut using a bowl chopper (Krag). The sample of shredded biomass was weighed and oven-dried at 105 °C until reaching constant mass. The moisture content was gravimetrically determined and the DM ratio was calculated (PN-EN 12880:2004).

The plants were mechanically harvested with a brush cutter (Stihl FS400 C, Germany) on 23 September 2016, 20 September 2017, and 11 October 2018. To estimate the fresh matter yield, the sorghum biomass was weighed just after harvest and the biomass production per hectare was extrapolated (Mg ha^{-1}). Harvesting losses were also included in calculations.

2.4. CO$_2$ Emission Determination and Carbon Footprint Calculation

The system boundaries for the carbon footprint calculation within the scope of this study are presented in Figure 1.

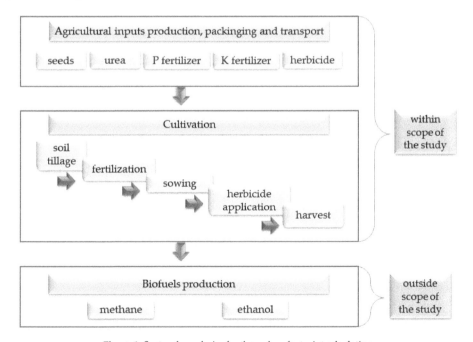

Figure 1. System boundaries for the carbon footprint calculation.

Calculations were performed based on Intergovernmental Panel on Climate Change methodology [18,40,41]. The main assumptions are given in Table 2. The quantification of GHG emissions was made according to ISO TS 14067 [42]. The freely available BioGrace Excel GHG calculation tool was used to estimate the C footprint of sorghum production [17]. Standard values containing conversion factors and LHV (lower heating values) from the database developed by IPCC were used for computing GHG emissions [43]. Other sources of emission factors are included in Table 3. The climate was classified as cold temperate and dry, and soil conditions were classified as sandy, in accordance with IPCC methodology. The environmental impact of different fertilization managements was assessed by estimating the spatial- and yield-scaled C footprint, expressed as kg CO_{2eq} ha^{-1} and kg CO_{2eq} Mg^{-1} DM produced, respectively. The assessment covers the major greenhouse gas emissions (CO_2, N_2O, and CH_4) generated during all major processes: from input materials for crop production, through the on-farm crop cultivation, to the field-gate. Analysis included both direct and indirect N_2O emissions. The following sources of direct N_2O emissions were incorporated into the analysis: N synthetic fertilizer, organic N applied as soil amendments (digestate and sewage sludge), and N in sorghum residues. Indirect N_2O was related to the atmospheric deposition of N volatilized from leaching and runoff, and CO_2 from urea fertilization.

Table 2. The main assumptions of the study.

Rule	Description
Scope of the study	Calculate the GHG emissions during sweet sorghum production for biofuels (methane and bioethanol) production.
System boundary	Farm stage—including external and on-farm greenhouse gas emissions.
Functional unit	1 ton of sorghum biomass.
Time reference	One growing season (as an average of three seasons).
Data collection—cultivation	The following agricultural operations were included: soil tillage, sowing, fertilization, herbicide application, and harvest.
Carbon footprint calculation:	
Calculator	BioGrace Excel GHG calculation tool [17]
Methods	IPCC 2006 [18,40,41]
Norm	ISO14067 [42]

The table structure was based on [21].

The GHG emissions were divided into two categories: (1) external emissions associated with production and transport of farm inputs, such as fertilizers, pesticides, and seeds and (2) on-site emissions including tractors and machinery fuel consumption during farm practices, and direct and indirect N_2O emissions. The GHG emissions from diesel oil consumption included all the operations of farm machinery used for various crop production activities, such as tillage, fertilizer and herbicide application, and sowing and harvesting. Annual CO_{2eq} emissions from urea application associated with the loss of CO_{2eq} during the industrial production process were calculated in the present study. Emissions related to soil C stock changes were included in the total C footprint of the farm. Quantification of GHG emissions was computed with emission factors according to the values shown in Table 3. Greenhouse gas emissions generated outside the farms (in wastewater treatment plants and during the biogas production process) were not considered. However, the analysis included emissions related to the application of sewage sludge and digestate and direct N_2O emissions due to N losses from soil fertilized with these kinds of organic amendments. According to IPCC methodology, it was assumed that there is no net accumulation of biomass C stocks. The change in biomass was not estimated, because for annual crops the increase in biomass stocks in a single year is equal to biomass losses from harvest and mortality in this year [41].

Conversion from N_2O-N emissions to N_2O was done by multiplication by 44/28. Emissions of CO_2, CH_4, and N_2O were quantified taking into account their 100-year global warming potentials (GWP), i.e., 1 for CO_2, 28 for CH_4, and 265 for N_2O. As recommended by IPCC, the most recent values of the 100-year time horizon GWP relative to CO_2 were used in this study, adapted from IPCC Fifth Assessment Report [44].

Table 3. Greenhouse gas emission factors for agricultural inputs and correlated coefficients used in the estimation in this study.

Description of Emission Factor	Unit	Default Value	References
Emission factor for N_2O emissions from N inputs	kg N_2O–N kg^{-1} N input	0.01	
$Frac_{GASF}$ fraction of synthetic fertilizer N that volatilizes as NH_3 and NO_x, kg N volatilized		10	
$Frac_{GASM}$ fraction of applied organic N fertilizer materials that volatilizes as NH_3 and NO_x, kg N volatilized		20	
Emission factor for N_2O emissions from atmospheric deposition of N on soils and water surfaces	%	1	[18]
$Frac_{LEACH-(H)}$ fraction of all N added to/mineralized in managed soils in regions where leaching/runoff occurs that is lost through leaching and runoff,		30	
Emission factor for N_2O emissions from N leaching and runoff		0.75	
Energy factor for urea production		20	
Fuel density (diesel)	kg m^{-3}	832	[17]
LHV (diesel) (1)	MJ kg^{-1}	43.1	
Emission factor for combustion of Diesel: CO_2 diesel		74100	
Emission factor for combustion of Diesel: CH_4 diesel	kg TJ^{-1}	4.15	[40]
Emission factor for combustion of Diesel: N_2O diesel		28.6	
Energy factor for mesotrione		691	
Energy factor for tetrabulazine and atriazine	MJ kg^{-1} a.i. (2)	208	[31]
Energy factor for metolachlor and metazachlor		388	
Energy factor for pesticide	kg CO_{2eq} MJ^{-1}	0.069	
Energy factor for P fertilizer production	kg CO_{2eq} kg^{-1} fertilizer	0.26	
Energy factor for K fertilizer production	kg CO_{2eq} kg^{-1} fertilizer	0.25	
Emission factor for sorghum seeds	g $CO_{2\,eq}$ kg^{-1}	0.86	[45]

(1) LHV lower heating value; (2) a.i. active ingredient.

2.5. Statistical Analyses

The normal distribution of the obtained data was checked with the Shapiro–Wilk test [46]. An analysis of variance (ANOVA) was performed to assess the statistical significance of the sorghum biomass yield and C footprint. The parameter averages were estimated across four plot replications. Treatment averages separation was carried out using the Tukey test at the probability level of $\alpha = 0.05$ [47]. The four tested fertilization managements were considered fixed factors and the four replications were entered as random effects [48]. The Levene test at P level < 0.05 was used for evaluation of the variance homogeneity. The Statistica software package (version 13.1 StatSoft, Poland) was used to carry out statistical analyses [49].

3. Results

3.1. Weather and Soil Conditions

In 2016, 2017, and 2018, the average temperatures during the sweet sorghum growing season were higher than the 30-year average: 0.8, 0.2, and 1.9 °C, respectively (Table 4). The period from August to September 2018 was characterized by the highest average temperature, which was 2.3 °C higher than the multiyear mean. In each experimental year, the total rainfall during the vegetation period was lower than the 30-year average: 74.6, 12.5, and 81.9 mm, respectively, for 2016, 2017, and 2018 (Table 4). Extremely adverse weather conditions with drought occurred at the beginning of the sorghum vegetation period in May 2016, when total precipitation was 90% lower than the 30year average. In 2017, both average temperature and precipitation were close to the long-term average temperature and rainfall sum.

Table 4. Monthly temperature and rainfall in the research area in the period of 2016–2018.

Month	$T_{average}$ (°C)				Rainfall (mm)			
	2016	2017	2018	Long-Term Average 1986–2015	2016	2017	2018	Long-Term Average 1986–2015
May	15.3	14.2	17.1	14.4	5.3	24.1	54.3	54.1
June	18.6	18.5	18.8	17.3	44.6	52.5	36.6	67.4
July	19.5	19.0	20.1	19.6	114.3	112.2	79.1	78.9
August	17.9	19.4	21.1	18.6	27.1	43.6	20.3	65.3
September	16.4	13.3	15.8	13.7	44.7	65.7	38.4	44.9
Average temperature or rainfall sum in the period from May to September	17.5	16.9	18.6	16.7	236.0	298.1	228.7	310.6

The experimental site is composed of loamy sand textured soils, originally classified as Brunic Arenosols soil (IUSS Working Group WRB, 2014). The soil was characterized by a thick (30–34 cm) humus layer and slightly acidic pH (6.0). The soil has a high content of nitrate and ammonium nitrogen, available phosphorus, and plant available form of potassium (Table 5). The soil is well-drained with the water table at 130 cm depth.

Table 5. Soil properties (topsoil layer 0–30 depth) at the study site.

Soil Texture			pH	NO_3-N	NH_4-N	$P_{available}$	$K_{available}$
%				$g\ kg^{-1}$		$mg\ kg^{-1}$	
sand: 87	silt: 5	clay: 8	6.0	0.79	0.55	337.5	154.0

3.2. Sorghum Biomass Yield

The results of sorghum biomass DM yield and C footprint are shown in Table 6. The analysis of variance showed that the DM yield production was not significantly affected by interaction between experimental factors considered in the study. The average sorghum DM yields ranged from 10.5 for non-fertilized SuperSile 20 to 23.6 Mg ha^{-1} for Sucrosorgo 506 fertilized with sewage sludge. In the case of each sorghum hybrid, the lowest biomass yield was produced by control plants. Hybrids produced significantly different biomass yields. Sucrosorgo 506 was characterized by the highest productivity in the study area. Dry matter yield of Sucrosorgo 506 was 35% higher than the yield of SuperSile 20, which produced the lowest yield of the four tested hybrids (Table 6). Both sewage sludge and digestate application significantly increased the yields across the hybrids, which were 45% and 28% higher than that for the control. The yields of sweet sorghum biomass were significantly different in the experimental period. In 2017, DM yield was 24.3% and 41.6% higher than in 2016 and 2018, respectively.

Table 6. Dry matter yield and carbon footprint of sorghum production as affected by different fertilization managements (pooled from 2016, 2017, and 2018).

Sorghum Hybrid	Fertilization Treatment	Dry Matter Yield Mg ha^{-1}	Spatial Carbon Footprint kg CO_{2eq} ha^{-1}	Yield-Scaled Carbon Footprint kg CO_{2eq}Mg^{-1}
Sucrosorgo 506	control	17.0	1731 [b]	88
	Urea	18.9	2742 [f]	130
	sewage sludge	23.6	2736 [f]	111
	digestate	19.0	2498 [def]	117
Rona 1	control	10.9	1414 [a]	96
	Urea	15.0	2528 [def]	141
	sewage sludge	15.3	2282 [cd]	126
	digestate	15.8	2303 [cd]	125
Goliath	control	12.2	1286 [a]	109
	Urea	17.9	2621 [ef]	135
	sewage sludge	19.6	2446 [cde]	118
	digestate	16.9	2340 [cde]	120
SuperSile 20	control	10.5	1412 [a]	96
	Urea	13.1	2472 [def]	147
	sewage sludge	14.8	2301 [cd]	124
	digestate	12.6	2180 [c]	131
	Average for factors Hybrid *			
Sucrosorgo 506		19.6 [c]	2427 [b]	111[a]
Rona 1		14.2 [ab]	2132 [a]	122[ab]
Goliath		16.6 [b]	2163 [ab]	120[ab]
SuperSile 20		12.7 [a]	2091 [a]	125[b]
	Fertilization treatment **			
control		12.6 [a]	1461 [a]	97[a]
urea		16.2 [b]	2590 [c]	138[c]
sewage sludge		18.3 [b]	2441 [b]	119[b]
digestate		16.1 [b]	2330 [b]	123[b]
	Average for years ***			
2016		15.3 [b]	2019 [a]	133[b]
2017		20.2 [c]	2282 [b]	114[a]
2018		11.8 [a]	2265 [b]	115[a]

Each value is the average of four replicates. [a,b,c] For interaction of factors analysis, each individual year and factor, averages with different letters in the same column are significantly different at $p < 0.05$ according to the results of ANOVA and Tukey's test. The columns without letters indicate that significant differences were not observed. * Values were averaged across four fertilization treatments and three years. ** Values were averaged across four hybrids and three years. *** Values were averaged across four hybrids and four fertilization treatments.

3.3. Carbon Footprint of Sorghum Per Area and Per Mg of Biomass

Averaged over the three years, a significant effect of both tested factors on spatial C footprint was reported in this study. The cultivation of sorghum without N fertilizer application resulted in the lowest CO_{2eq} per unit of area and per unit of biomass yield; in particular, in the case of the three hybrids: Rona 1, Goliath, and SuperSile 20. Irrespective of fertilization treatments, the highest yielding hybrid, Sucrosorgo 506, emitted the highest amount of GHG per hectare. Emissions from field production of this hybrid were 16% higher than those from cultivation of SuperSile 20, which was characterized by the lowest yield and the lowest area-scaled C footprint. The spatial C footprint was significantly affected by fertilization management systems. The application of waste products caused a decrease in GHG emissions compared to conventional fertilization. The application of sewage sludge and digestate resulted in 6% and 10% lower CO_{2eq} emissions per ha, respectively, compared with the use of urea. The control sorghum plants produced the lowest amount of GHG. The spatial C footprint varied greatly during the period of the experiment. In 2017 and 2018, this was at a comparable level, which was on average 12.5% higher than in 2016.

The yield-scaled C footprint expressed as a CO_{2eq} per Mg DM of yield produced presented a quite narrow range from 88 to 147 kg CO_{2eq} Mg^{-1} DM, without significant differences between the interactions of tested factors. Greenhouse gas emissions per Mg of biomass were the highest from

SuperSile 20 hybrid cultivation, which was characterized by the lowest biomass yield. Across the hybrids, the effect of fertilization treatment was the same as for spatial C footprint, with the highest value for urea application. Application of sewage sludge and digestate provided, respectively, 14% and 11% lower emissions of GHG than from the use of urea. Across all hybrids and fertilization treatments, differences between years were also significant. In 2016, sorghum cultivation emitted higher amounts of GHG per Mg of produced biomass compared to 2017 and 2018.

3.4. Structure of Inputs Share of Carbon Footprint

Across all the hybrids and years, the share of varied inputs in the C footprint is presented for each fertilization treatment in Figure 2. When the emissions were averaged across all hybrids, the largest contributor to the total amount of GHG emissions was the combination of direct and indirect N_2O emissions, which ranged from 56% to 63% of the total emissions and from 10% to 17%, respectively, for direct and indirect N_2O emissions. An increase in direct N_2O emissions was observed when sorghum was fertilized with sewage sludge and digestate (Figure 2C,D). When urea was applied as the N source, it was responsible for 13% of the total GHG emission and this was the second largest contributor to the C footprint. Diesel combustion during various farming operations (such as soil tillage, sowing, herbicide spraying, fertilizer application, harvesting, etc.) was responsible for relatively high GHG emissions and its contribution to the C footprint of sorghum production was estimated at 13–22%. Carbon dioxide equivalents emissions related to the use of P and K fertilizers, seeds and herbicide had the lowest contribution to the total GHG emission and on average they all accounted for from 5% to 9% (Figure 2A–D).

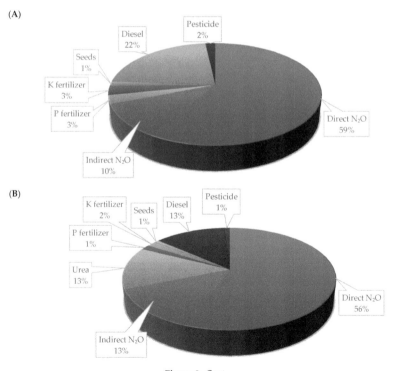

Figure 2. *Cont.*

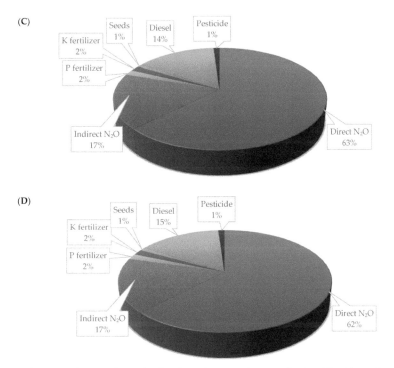

Figure 2. Structure of inputs share of carbon footprint in sorghum production (**A**) without nitrogen fertilization, fertilized with (**B**) urea, (**C**) sewage sludge, and (**D**) digestate (pooled from 2016, 2017, and 2018).

3.5. External and On-Site Emissions

The sources of external GHG emissions are shown in Figure 3. In the analysis, the following sources of external emission were included: manufacture, transportation, storage, and delivery of agricultural inputs to the farm gate. The use of N fertilizer was responsible for 54% of external emissions and dominated in terms of the external GHG emissions. Consequently, the external emissions from sorghum fertilized with urea were 2.2 times higher than emissions from treatments fertilized with sewage sludge and digestate (Figure 3).

As an average of four hybrids and three years, on-site emissions made a significantly higher contribution to the total GHG emission, because these were from 8 to 17 times higher than external emissions (Figures 3 and 4). Direct N_2O was responsible for the largest amount of emitted GHG in each fertilization treatment. Averaged over three years, diesel consumption was the second largest contributor to the C footprint at the farm level. On-farm emissions from unfertilized sorghum were 27%, 19%, and 22% lower compared to those from sorghum fertilized in a conventional way and by application of sewage sludge and digestate as soil amendments, respectively. On-site emissions were similar between fertilization treatments that used synthetic N or biosolids as a source of nutrients (Figure 4).

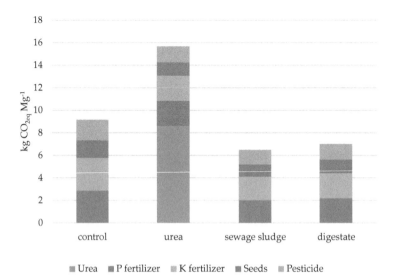

Figure 3. External emissions of CO_2 related to agricultural inputs, such as N, P, and K fertilizers, seeds, and pesticide, as affected by different fertilization managements. Values correspond to the annual average for the 2016–2018 period.

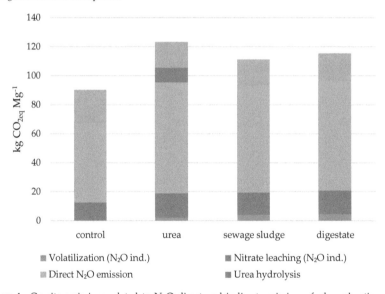

Figure 4. On-site emissions related to N_2O direct and indirect emissions, fuel combustion, and urea hydrolysis from sorghum production as affected by different fertilization managements. Values correspond to the annual average for the 2016–2018 period.

4. Discussion

4.1. Biomass Yield

The DM biomass yield of Rona 1 fertilized with urea recorded in this study (15.0 Mg ha^{-1}) was similar to that reported in a previous study carried out under the same environmental conditions (15.2 Mg ha^{-1}) [36]. In previous research at the same experimental site, a lower DM yield for sweet sorghum hybrid Sucrosorgo 304 fertilized with urea compared to the yield of Sucrosorgo 506 was reported [49]. Irrespective of the hybrids, application of bio-based waste products provided a biomass yield statistically comparable to that for crops fertilized in a conventional way with urea. These observations are in line with the findings of Kołodziej et al. [23], who reported that application of sewage sludge enhanced yields of Sucrosorgo 506 and Rona 1. Similarly, Akdeniz et al. [50] recorded an increase in sorghum DM yield as a response to sewage sludge application. Our findings are congruent with the results of Verdi et al. [51], who found no differences between the yield of crops fertilized with digestate and urea. In another study conducted in southwestern Germany, it was found that digestate could be an adequate substitute for mineral fertilizer in sweet sorghum production [52]. The same reaction of sorghum plants to biosolids application was noted by Sigurnjak et al. [53] in a study carried out in the Czech Republic under similar weather conditions. Digestate can be considered a synthetic N substitute without crop yield losses [54]. The biomass yields were significantly different across the years of the experiment. This was probably associated with the varied weather conditions.

4.2. Carbon Footprint

Results indicated that greenhouse gas emissions varied considerably between the fertilization treatments. Both the CO_{2eq} emitted to produce a metric ton of biomass and emissions per area unit decreased when sewage sludge and digestate were applied. Styles and Jones [55] reported that production of miscanthus biomass for energy purposes resulted in GHG emissions of 1938 kg CO_{2eq} per hectare. This lower value can be associated with the lower N demand of miscanthus [33].

The application of synthetic fertilizer is the main source of external GHG emissions from sorghum production. Similar results were obtained in a study conducted by Plaza-Bonilla et al. [11], who reported the great impact of N fertilization on external emissions. Most of the C footprint is associated with N fertilizer production and use [19]. Storlien et al. [9] reported a significant impact of N fertilization on CO_2 and N_2O emissions from sweet sorghum, especially at the beginning of the growing season in each year of the experiment.

Lower emissions from the production of sorghum using biosolids were associated with the lower reliance on the external input of synthetic fertilizer. Considering the CO_{2eq} quantity emitted into the atmosphere for synthetic fertilizers production, partial or total fertilization with digestate provided lower CO_{2eq} emissions [25]. Application of digestate had a relatively lower impact on the emissions of CO_2 and CH_4 compared to urea [56]. Cumulative N emissions via volatilization showed that digestate could be a promising method of sustainable fertilization management to decrease N losses [51]. However, these research projects pointed out differences between gases emitted by the two kinds of fertilizers: digestate emitted 23% more N_2O than urea, but urea emitted 66% more ammonia than digestate [51]. These results are congruent with the findings of Dendooven et al. [26], who reported that emissions of CH_4 and CO_2 were not affected by fertilization treatments; however, digestate application increased emissions of N_2O. It was revealed that combining fresh and more stabilized sewage sludge enabled a decrease in N_2O emissions [44,57]. The emissions intensity of GHG from digestate amended soils was lower compared to the use of untreated manure and was at a similar level those for synthetic fertilizer. However, the researchers pointed out that the agronomic and environmental results related to the impact on crop yield and C footprint cannot simply be predicted on this basis; rather, specific soil and digestate physio-chemical characteristics should also be taken into account [57].

Diesel combustion during farming operations is a significant contributor to global warming potential [12]. Pesticide application accounts for the lowest GHG contribution and this is associated with the low demand of sorghum for plant protection chemicals used for weed, disease, and insect control. Only a low rate of herbicide is needed in sorghum cultivation, which is a new crop in the Central European region and does not yet have natural enemies. Findings of the present study are in line with Plaza-Bonilla et al. [11], who found that emissions related to pesticides represented only 1%

of external emissions, as an average of the tested hybrids and years. Sweet sorghum can be recognized as a high-yielding biofuel feedstock with minimal impact on net GHG emissions [58].

5. Conclusions

This study was performed because there is limited information on the allocation of agricultural residues for sweet sorghum as well as for digestate, which is a sub-product of anaerobic digestion. Application of bio-based by-products (sewage sludge and digestate) provided a sorghum biomass yield close to that obtained when conventional synthetic fertilizer was used. Combined direct and indirect CO_{2eq} emissions on the basis of both per unit area and per unit of biomass yield were lower when bio-based waste products were used compared with the application of urea. The present study indicated that the lower GHG emissions resulted from the reduced reliance on synthetic N fertilizers due to their replacement by alternative nutrient sources, such as sewage sludge and digestate. It can be concluded that this fertilization practice can be considered a promising sustainable strategy for low carbon agriculture, which allows the recycling of nitrogen and other nutrients as an element of the circular economy. Biosolids should be recommended for providing sustainable sorghum production as a feedstock for bioenergy to mitigate GHG emissions and global climate change processes. Further research is needed to confirm the suitability of the alternative fertilization management proposed in the present study. This future work should be focused on comparing results based on other available calculation tools. There is a great need to develop local emission factors, which will provide better characteristics of national conditions. The system boundaries can be extended to the production process of bioethanol and methane from sorghum. Moreover, data thus obtained could be compared with results from direct measurements of GHG emissions from soil using chambers placed in the field.

Author Contributions: Conceptualization, L.G. and J.S.; methodology, L.G. and J.S.; validation, J.S.; formal analysis, J.S. and L.G.; investigation, L.G. and J.S.; resources, L.G. and J.S.; data curation, L.G.; writing—original draft preparation, L.G.; writing—review and editing, L.G. and J.S.; visualization, L.G.; supervision, J.S.

Funding: This research received no external funding.

Conflicts of Interest: The authors declare no conflict of interest.

References

1. Usman, M.; Ibrahim, F.; Oyetola, S.O. Sustainable agriculture in relation to problems of soil degradation and how to amend such soils for optimum crop production in Nigeria. *Int. J. Res. Agric. Food Sci.* **2018**, *4*, 1–17.
2. Brodt, S.; Six, J.; Feenstra, G.; Ingels, C.; Campbell, D. Sustainable Agriculture. *Nat. Educ. Knowl.* **2011**, *3*, 1.
3. Vergé, X.P.C.; De Kimpe, C.; Desjardins, R.L. Agricultural production, greenhouse gas emissions and mitigation potential. *Agric. Meteorol.* **2007**, *142*, 255–269. [CrossRef]
4. IPCC. *Climate Change 2014: Synthesis Report. Contribution of Working Groups I, II and III to the Fifth Assessment Report of the Intergovernmental Panel on Climate Change*; Core Writing Team, Pachauri, R.K., Meyer, L.A., Eds.; IPCC: Geneva, Switzerland, 2014.
5. EEA. *Trends and Projections in Europe 2018. Tracking Progress towards Europe's Climate and Energy Targets*; European Environment Agency report 16/2018; Publications Office of the European Union: Luxembourg, 2018.
6. Muñoz, C.; Paulino, L.; Monreal, C.; Zagal, E. Greenhouse Gas (CO_2 AND N_2O) Emissions from Soils: A Review. *Chil. J. Agric. Res.* **2010**, *70*, 485–497. [CrossRef]
7. Frank, S.; Havlík, P.; Soussana, J.F.; Levesque, A.; Valin, H.; Wollenberg, E.; Kleinwechter, U.; Fricko, O.; Gusti, M.; Herrero, M.; et al. Reducing greenhouse gas emissions in agriculture without compromising food security? *Environ. Res. Lett.* **2017**, *12*, 105004. [CrossRef]
8. Sosulski, T.; Rutkowska, B.; Szczepaniak, J.; Szulc, W.; Skowrońska, M. Impact of reduced tillage on CO_2 emission from soil under maize cultivation. *Soil Tillage Res.* **2018**, *180*, 21–28. [CrossRef]
9. Storlien, J.O.; Hons, F.M.; Wight, J.P.; Heilman, J.L. Carbon Dioxide and nitrous oxide emissions impacted by bioenergy sorghum management. *Soil Sci. Soc. Am. J.* **2014**, *78*, 1694–1706. [CrossRef]

10. Peter, C.; Helming, K.; Nendel, C. Do greenhouse gas emission calculations from energy crop cultivation reflect actual agricultural management practices? —A review of carbon footprint calculators. *Renew. Sustain. Energy Rev.* **2017**, *67*, 461–476. [CrossRef]

11. Plaza-Bonilla, D.; Nogué-Serra, I.; Raffaillac, D.; Cantero-Martínez, C.; Justes, É. Carbon footprint of cropping systems with grain legumes and cover crops: A case-study in SW France. *Agric. Syst.* **2018**, *167*, 92–102. [CrossRef]

12. Yadav, G.S.; Das, A.; Lal, R.; Babu, S.; Meena, R.S.; Saha, P.; Singh, R.; Datta, M. Energy budget and carbon footprint in a no-till and mulch based rice–mustard cropping system. *J. Clean. Prod.* **2018**, *191*, 144–157. [CrossRef]

13. Bos, J.F.F.P.; De Haan, J.; Sukkel, W.; Schils, R.L.M. Energy use and greenhouse gas emissions in organic and conventional farming systems in the Netherlands. *NJAS Wagening. J. Life Sci.* **2014**, *68*, 61–70. [CrossRef]

14. European Parliament and Council. Directive 2009/30/EC of 23 April 2009 as Regards the Specification of Petrol, Diesel and Gas-Oil and Introducing A Mechanism to Monitor and Reduce Greenhouse Gas Emissions. 2009. Available online: http://eurlex.europa.eu/LexUriServ/LexUriS (accessed on 6 February 2019).

15. European Parliament and Council. Directive 2009/28/EC of 23 April 2009 on the Promotion of the Use of Energy from Renewable Sources. 2009. Available online: http://eurlex.europa.eu/LexUriServ/LexUriServ.do?uri=OJ:L:2009:140:0016:0062:EN:PDF (accessed on 12 January 2019).

16. Hennecke, A.M.; Faist, M.; Reinhardt, J.; Junquera, V.; Neeft, J.; Fehrenbach, H. Biofuel greenhouse gas calculations under the European Renewable Energy Directive—A comparison of the BioGrace tool vs. the tool of the Roundtable on Sustainable Biofuels. *Appl. Energy* **2013**, *102*, 55–62. [CrossRef]

17. BioGrace Software. Available online: http://www.biograce.net (accessed on 23 February 2019).

18. IPCC. Chapter 11: N_2O Emissions from Managed Soils, and CO_2 Emissions from Lime and Urea Application. In *IPCC Guidelines for National Greenhouse Gas Inventories*; Institute for Global Environmental Strategies (IGES): Kanagawa, Japan, 2006; pp. 11.1–11.54.

19. Shen, Y.; Sui, P.; Huang, J.; Wang, D.; Whalen, J.K.; Chen, Y. Global warming potential from maize and maize-soybean as affected by nitrogen fertilizer and cropping practices in the North China Plain. *Field Crops Res.* **2018**, *225*, 117–127. [CrossRef]

20. Camargo, G.G.T.; Ryan, M.R.; Richard, T.L. Energy Use and Greenhouse Gas Emissions from Crop Production Using the Farm Energy Analysis Tool. *Bioscience* **2013**, *63*, 263–273. [CrossRef]

21. Bartocci, P.; Fantozzi, P.; Fantozzi, F. Environmental impact of Sagrantino and Grechetto grapes cultivation for wine and vinegar production in central Italy. *J. Clean. Prod.* **2017**, *140*, 569–580. [CrossRef]

22. COM. *Proposal for a Regulation of the European Parliament and of the Council Laying Down Rules on the Making Available on the Market of CE Marked Fertilising Products and Amending Regulations (EC) No 1*; 157 final, 17.3.2016 2016/0084 (COD); COM: Brussels, Belgium, 2016.

23. Kołodziej, B.; Antonkiewicz, J.; Stachyra, M.; Bielińska, E.J.; Wiśniewski, J.; Luchowska, K.; Kwiatkowski, C. Use of sewage sludge in bioenergy production—A case study on the effects on sorghum biomass production. *Eur. J. Agron.* **2015**, *69*, 63–74. [CrossRef]

24. Ho, A.; Ijaz, U.Z.; Janssens, T.K.S.; Ruijs, R.; Kim, S.Y.; de Boer, W.; Termorshuizen, A.; van der Putten, W.H.; Bodelier, P.L.E. Effects of bio-based residue amendments on greenhouse gas emission from agricultural soil are stronger than effects of soil type with different microbial community composition. *Gcb Bioenergy* **2017**, *9*, 1707–1720. [CrossRef]

25. Maucieri, C.; Nicoletto, C.; Caruso, C.; Sambo, P.; Borin, M. Effects of digestate solid fraction fertilisation on yield and soil carbon dioxide emission in a horticulture succession. *Ital. J. Agron.* **2017**, *12*, 116–123. [CrossRef]

26. Dendooven, L.; Fernández-Luqueño, F.; Paredes-López, O.; Hernández, G.; Pampillón-González, L.; Franco-Hernández, O.; Luna-Guido, M.; Ruíz-Valdiviezo, V.M. Greenhouse Gas Emissions and Growth of Wheat Cultivated in Soil Amended with Digestate from Biogas Production. *Pedosphere* **2017**, *27*, 318–327. [CrossRef]

27. Juárez-Rodríguez, J.; Fernández-Luqueño, F.; Conde, E.; Reyes-Varela, V.; Cervantes-Santiago, F.; Botello-Alvarez, E.; Cárdenas-Manríquez, M.; Dendooven, L. Greenhouse gas emissions from an alkaline saline soil cultivated with maize (*Zea mays* L.) and amended with anaerobically digested cow manure: A greenhouse experiment. *J. Plant Nutr.* **2012**, *35*, 511–523. [CrossRef]

28. Walsh, J.J.; Jones, D.L.; Edwards-Jones, G.; Williams, A.P. Replacing inorganic fertilizer with anaerobic digestate may maintain agricultural productivity at less environmental cost. *J. Plant Nutr. Soil Sci.* **2012**, *175*, 840–845. [CrossRef]

29. Pezzolla, D.; Bol, R.; Gigliotti, G.; Sawamoto, T.; López, A.L.; Cardenas, L.; Chadwick, D. Greenhouse gas (GHG) emissions from soils amended with digestate derived from anaerobic treatment of food waste. *Rapid Commun. Mass Spectrom.* **2012**, *26*, 2422–2430. [CrossRef]

30. Allen, B.; Kretschmer, B.; Baldock, D.; Menadue, H.; Nanni, S.; Tucker, G. *Space for Energy Crops—Assessing the Potential Contribution to Europe's Energy Future*; Report Produced for BirdLife Europe; European Environmental Bureau and Transport & Environ: Brussels, Belgium, 2014.

31. Brentrup, F.; Palliere, C. Energy efficiency and greenhouse gas emissions in European nitrogen fertilizer production and use. *Fertil. Eur.* **2008**, *639*, 25.

32. Chmielewska, J.; Sowiński, J.; Foszczyńska, B.; Szydełko-Rabska, E.; Kawa-Rygielska, J. Production of bioethanol from sweet sorghum juices with varying content of mineral compounds. *Przem. Chem.* **2014**, *93*, 999–1001.

33. Davis, S.C.; Parton, W.J.; Del Grosso, S.J.; Keough, C.; Marx, E.; Adler, P.R.; Delucia, E.H. Impact of second-generation biofuel agriculture on greenhouse-gas emissions in the corn-growing regions of the US. *Front. Ecol. Environ.* **2012**, *10*, 69–74. [CrossRef]

34. Dubicki, A.; Dubicka, M.; Szymanowski, M. Klimat Wrocławia. In *Środowisko Wrocławia-Informator*; Smolnicki, K., Szykasiuk, M., Eds.; Dolnośląska Fundacja Ekorozwoju: Wrocław, Poland, 2002; pp. 9–20. (In Polish)

35. Szczepaniak, W. *Metody Instrumentalne w Analizie Chemicznej*; PWN: Warszawa, Poland, 2005. (In Polish)

36. Sowiński, J.; Szydełko-Rabska, E. Comparison of sorghum type yielding in Polish conditions. *Ann. Univ. Mariae Curie-Skłodowska Lub. Pol. Sect. E* **2013**, *LXVIII*, 30–40. (In Polish)

37. Hybrid Characteristic. Available online: https://www.gabonakutato.hu/en/our-seeds/sorghum/silage-sorghum/rona-1 (accessed on 13 January 2019).

38. Hybrid Characteristic. Available online: https://www.caussade-nasiona.pl/sorgo-trawa-sudanska/ (accessed on 17 January 2019).

39. Hybrid Characteristic. Available online: https://www.benelux.saaten-union.com/index.cfm?m=varieties&p=325,1683,html (accessed on 17 January 2019).

40. IPCC (Intergovernmental Panel on Climate Change). Chapter 3. Mobile combustion. In *Guidelines for National Greenhouse Gas Inventories*; Institute for Global Environmental Strategies (IGES): Kanagawa, Japan, 2006; Volume 2, Energy. Available online: https://www.ipcc-nggip.iges.or.jp/public/2006gl/vol2.html (accessed on 23 February 2019).

41. IPCC (Intergovernmental Panel on Climate Change). Cropland. In *IPCC Guidelines for National Greenhouse Gas Inventories. Agriculture, Forestry and Other Land Use*; Institute for Global Environmental Strategies (IGES): Kanagawa, Japan, 2006; Volume 4. Available online: http://www.ipcc-nggip.iges.or.jp/public/2006gl/vol4.h (accessed on 15 January 2019).

42. ISO. *ISO/TS 14067: Greenhouse Gases e Carbon Footprint of Products e Requirements and Guidelines for Quantification and Communication*; International Organization for Standardization: Geneva, Switzerland, 2013.

43. BioGrace List of Standard Values, Version 4. Available online: http://www.biograce.net/content/ghgcalculationtools/standardvalues (accessed on 11 January 2019).

44. Nielsen, S.; Bruun, S.; Bekiaris, G.; Gómez-Muñoz, B.; Larsen, J.D.; Jensen, L.S.; Scheutz, C. Nitrogen mineralisation and greenhouse gas emission from the soil application of sludge from reed bed mineralisation systems. *J. Environ. Manag.* **2017**, *203*, 59–67. [CrossRef]

45. West, T.O.; Marland, G. A synthesis of carbon sequestration, carbon emissions, and net carbon flux in agriculture: Comparing tillage practices in the United States. *Agric. Ecosyst. Environ.* **2002**, *91*, 217–232. [CrossRef]

46. Ghasemi, A.; Zahediasl, S. Normality tests for statistical analysis: A guide for non-statisticians. *Int. J. Endocrinol. Metab.* **2012**, *10*, 486. [CrossRef]

47. Pazderu, K.; Hodoval, J.; Urban, J.; Pulkrabek, J.; Pacuta, V.; Adamcik, J. The influence of sweet sorghum crop stand arrangement on biomass and biogas production. *Plant Soil Environ.* **2014**, *60*, 433–438. [CrossRef]

48. Maharjan, B.; Rosen, C.J.; Lamb, J.A.; Venterea, R.T. Corn response to nitrogen management under fully-irrigated vs. water-stressed conditions. *Agron. J.* **2016**, *108*, 2089. [CrossRef]

49. Sowiński, J.; Głąb, L. The effect of nitrogen fertilization management on yield and nitrate contents in sorghum biomass and bagasse. *Field Crops Res.* **2018**, *227*, 132–143. [CrossRef]
50. Akdeniz, H.; Yilmaz, I.; Bozkurt, M.A.; Keskin, B. The effects of sewage sludge and nitrogen applications on grain sorghum grown (*Sorghum vulgare* L.) in Van-Turkey. *Pol. J. Environ. Stud.* **2006**, *15*, 19–26.
51. Verdi, L.; Kuikman, P.J.; Orlandini, S.; Mancini, M.; Napoli, M.; Dalla Marta, A. Does the use of digestate to replace mineral fertilizers have less emissions of N_2O and NH3? *Agric. Meteorol.* **2019**, *269–270*, 112–118. [CrossRef]
52. Formowitz, B.; Fritz, M. Biogas Diegstate as Organic Fertilizer in Different Crop Rotations. In Proceedings of the 18th European Biomass Conference and Exibition, Lyon, France, 3–7 May 2010; pp. 224–229.
53. Jaša, S.; Badalíková, B.; Červinka, J. Influence of Digestate on Physical Properties of Soil in Zd Budišov. *Acta Univ. Agric. Silvic. Mendel. Brun.* **2019**, *67*, 75–83. [CrossRef]
54. Sigurnjak, I.; Ryckaert, B.; Tack, F.M.G.; Ghekiere, G.; Meers, E.; Vaneeckhaute, C.; Michels, E. Fertilizer performance of liquid fraction of digestate as synthetic nitrogen substitute in silage maize cultivation for three consecutive years. *Sci. Total Environ.* **2017**, *599–600*, 1885–1894. [CrossRef]
55. Styles, D.; Jones, M.B. Energy crops in Ireland: Quantifying the potential life-cycle greenhouse gas reductions of energy-crop electricity. *Biomass Bioenergy* **2007**. [CrossRef]
56. Czubaszek, R.; Wysocka-Czubaszek, A. Emissions of carbon dioxide and methane from fields fertilized with digestate from an agricultural biogas plant. *Int. Agrophysics* **2018**, *32*, 29–37. [CrossRef]
57. Baral, K.R.; Labouriau, R.; Olesen, J.E.; Petersen, S.O. Nitrous oxide emissions and nitrogen use efficiency of manure and digestates applied to spring barley. *Agric. Ecosyst. Environ.* **2017**, *239*, 188–198. [CrossRef]
58. Storlien, J.O. The Carbon Footprint of Bioenergy Sorghum Production in Central Texas: Production Implications on Greenhouse Gas Emissions, Carbon Cycling, and Life Cycle Analysis. Ph.D. Thesis, Texas A&M University, College Station, TX, USA, 2013.

Article

Mapping the Environmental Cost of a Typical Citrus-Producing County in China: Hotspot and Optimization

Min Yang [1], Quan Long [2], Wenli Li [1], Zhichao Wang [1], Xinhua He [1], Jie Wang [1], Xiaozhong Wang [1], Huaye Xiong [1], Chaoyi Guo [1], Guancheng Zhang [3], Bin Luo [4], Jun Qiu [4], Xinping Chen [1,5,6,7], Fusuo Zhang [2,6], Xiaojun Shi [1,5,6,7] and Yueqiang Zhang [1,5,6,7,*]

[1] College of Resources and Environment, Southwest University, Chongqing 400716, China; yangmin0827@email.swu.edu.cn (M.Y.); zmylwl20@email.swu.edu.cn (W.L.); w15353152989@email.swu.edu.cn (Z.W); xinhua.he@uwa.edu.au (X.H.); mutouyu@swu.edu.cn (J.W.); wxz20181707@swu.edu.cn (X.W.); woaisesero@email.swu.edu.cn (H.X.); gcy1994@email.swu.edu.cn (C.G.); chenxp2017@swu.edu.cn (X.C.); shixj@swu.edu.cn (X.S.)
[2] College of Resources and Environmental Sciences, China Agricultural University, Beijing 100193, China; nongyuanlq@163.com (Q.L.); zhangfs@cau.edu.cn (F.Z.)
[3] Yunnan Yuntianhua Co., Ltd., Kunming 650228, China; zgc18387389540@163.com
[4] Danling County Agricultural and Rural Bureau, Meishan 620200, China; 108_818@163.com (B.L.); yunjun9842@126.com (J.Q.)
[5] Academy of Agricultural Sciences, Southwest University, Chongqing 400716, China
[6] Interdisciplinary Research Center for Agriculture Green Development in Yangtze River Basin, Southwest University, Chongqing 400716, China
[7] State Cultivation Base of Eco-agriculture for Southwest Mountainous Land, Southwest University, Chongqing 400716, China
[*] Correspondence: zhangyq82@swu.edu.cn; Tel.: +86-1992-335-8234

Received: 10 February 2020; Accepted: 25 February 2020; Published: 28 February 2020

Abstract: The environmental sustainability of the largest citrus plantation globally is facing a great challenge in China. Further, there is a lack of quantitative, regional hotspot studies. In this study, the life cycle assessment (LCA) was used to quantify the environmental cost of citrus production based on 155 farmers' surveys from typical citrus orchards in Danling County, southwest China, which produced 0.65% of the country's total citrus production. The results showed that the average values of environmental risk indicated by global warming potential (GWP), acidification potential (AP), and eutrophication potential (EP) were 11,665 kg CO_2-eq ha^{-1}, 184 kg SO_2-eq ha^{-1}, and 110 kg PO_4-eq ha^{-1}, respectively. The production and utilization of fertilizer ranked the first contribution to the environmental impacts among all the environmental impacts, which contributed 92.4–95.1%, 89.4–89.8%, and 97.8–97.9% to global warming potential, acidification potential, and eutrophication potential, respectively. Specific to the contribution of fertilizers to environmental costs, the production and utilization of nitrogen (N) fertilizer accounted for more than 95% of the total environmental costs. Thus, the spatial distribution of environmental costs in this county was well matched with that of N input. Compared with the average values of investigated 155 orchards, the high yield and high N use efficiency (HH) orchard group with younger and better educated owners achieved a higher citrus yield and N use efficiency with less fertilizer input and lower environmental costs. Five field experiments conducted by local government and Danling Science and Technology Backyard were used to further certify the reduction potential of environment costs. These field results showed that the local recommendation (LR) treatment increased citrus yield and N use efficiency by 1.9–49.5% and 38.0–116%, respectively, whereas decreased environmental costs by 21.2–35.2% when compared with the local farmer practice in the HH orchard group. These results demonstrated that an optimum nutrient management based on the local field recommendation in citrus-producing areas is crucial for achieving a win-win target of productivity and environmental sustainability in China and other, similar countries.

Sustainability **2020**, *12*, 1827

Keywords: environmental cost; life cycle assessment; citrus; nitrogen; optimization

---

1. Introduction

Citrus is the top fruit crop with the largest cultivation area and highest production in the world [1]. China is the largest citrus producer with 28.3% of global cropping areas and 26.8% of global production [1]. However, the environmental sustainability of citrus production in China is facing a great challenge, because of its low yield but high inputs per unit area in agricultural materials [2]. Over the past 20 years, the citrus yield in China (15.0 t ha^{-1}) has increased by 2.1 times, but is still lower than global average yield and less than half of that in South Africa (32.9 t ha^{-1}) [1,3]. Driven by economic interests, fertilization has become the major approach to increase citrus yield in China. According to nationwide survey in China, the average application rate of nitrogen (N) fertilizer in citrus orchards is as high as 500 kg ha^{-1} [4,5], which is substantially higher than that in advanced citrus-producing countries such as Brazil (200 kg ha^{-1}) and USA (150–200 kg ha^{-1}) [6,7]. Such a high rate of N and other nutrient fertilization has far exceeded the requirement of citrus trees in China's citrus orchards with low yield. This would also result in lower fertilizer use efficiency [4] and projected serious nutrient loss to environment, which will inevitably lead to environmental risks [8]. At present, global warming, acidification and eutrophication are becoming major challenges in intensive agriculture systems in China, which were mainly driven by excessive nutrient inputs [9–11]. Without doubt, the citrus producing system in China is in conflict with the goal of zero growth of chemical fertilizer and the green development of the citrus industry. Therefore, it is urgent to assess the environmental cost of typical citrus producing region in China, and to close the citrus yield gap through nutrient management for sustainable intensification.

In order to quantify environmental sustainability, methods have been developed and used in the agriculture section to identify the negative externalities produced during agricultural processes. These methods include healthy farmland system assessment framework (HFSAF), life cycle assessment (LCA), sustainability assessment of farming and the environment (SAFE), multi-temporal remote sensing images, and so on [12–15]. Among them, the LCA has been a preferential method to quantify the potential environmental impacts during agricultural production and processing and has been applied to agricultural systems since the 1990s [16,17].

Although many aspects of environmental costs in cereal and annual crop productions have already been investigated, the LCA of environmental indicators in the perennial fruit crops is still rare, mainly due to lack of methodological standardization [18,19]. For citrus research, environmental assessments of integrated citrus productions in Spain were firstly evaluated by LCA to detect the hot spots, including agrochemicals, energy, and agricultural practices [20]. Thereafter, the environmental assessment of citrus orchards by LCA has been conducted in Italy, Korea, Brazil, and Iran [18,21]. The LCA has been also used to compare the greenhouse gas (GHG) emission among orchards of citrus and other major fruit crops in China [22]. In similar studies, environmental costs have been quantified and major contributors have been sorted out [21,23,24]. Further, some optimal scenarios with fewer environmental impacts have been also suggested, although only a few studies are widely applied [25].

It is also known that the environmental costs of crop production systems vary greatly among different regions due to differences in soil types, climate condition, and management practices [18]. China's citrus production is dominated by small-scale farmers, and there are great differences in management practices among regions [4]. Thus, it is expected that the contribution of different factors to environmental costs is different with high variability. By learning from farmer practices, farmer group with high yield and high nutrient efficiency represent the realizable potential to reduce environmental cost in studied region [26–28]. Therefore, it is necessary to quantify the environmental cost and to detect their hotspots in typical citrus-producing areas in China, so as to provide scientific recommendations for regionally environmental and agricultural management.

As a basic administrative region, study at a county level can provide insight into the overall status of citrus production in China. Danling County, locating in Sichuan province of southwest China, is one of top 30 citrus producing counties in China [2]. Through 30 years' cultivation and improvement, citrus orchards (10,667 ha) with the dominant variety of *Shiranuhi* have been developed in Danling County. This cropping area accounts for 0.41% of the citrus in China and 0.11% of global citrus. Whereas the total production of citrus in Danling county is 266,000 tons and their output value reaches 2.66 billion Yuan, accounting for 2.1% of the national output value of citrus [1,3]. Preliminary surveys conducted by the Danling Science and Technology Backyards have showed that this typical high-yield and high-benefit citrus production system also feeds more than 500 agricultural material distributors. However, the environmental sustainability of citrus production in Danling County remains unknown. Therefore, taking the Danling County as an example, the aims of this study were: 1) to quantify and locate the environmental cost of citrus production by a county-level farmer investigation and the LCA method; 2) to assess the achievable potential to reduce the environmental impact by a farmer grouping strategy; and 3) to test the further potential of reducing environmental cost by addressing the detected prominent problems through field demonstrations. This study would be an example for other citrus-producing counties in China or other countries to identify and manage the hotspots of environmental impacts.

2. Materials and Methods

2.1. Studied Region

The study was conducted in Danling County of Sichuan Province, located in the southwest edge of Chengdu Plain, southwest China (103.23°~103.58°E, 29.87°~30.13°N). This region has a subtropical humid climate zone with mild climate. The annual average temperature is 16.7 °C. The annual average precipitation is around 1,158mm (932.7mm between May and September). The primary soil type is yellow soil, and the soil texture is clay. Citrus is the main economic crop in this area, accounting for 58% of the county's cultivated area (Supplementary Figure S1a). This research was conducted by the Danling Science and Technology Backyard (STB) (Supplementary Figure S1b), which was jointly established by Southwest University, China Agricultural University, Citrus Research Institute of the Chinese Academy of Agricultural Sciences, and the Danling County People's Government in January 2017. It involves in agricultural scientists living in villages among farmers for advancing participatory innovation and technology transfer and the Danling citrus industry to improve quality, efficiency, and the green development of agriculture.

2.2. Data Collection and Processing

2.2.1. Data Collection

According to the method of a farmer survey, 3–4 villages were randomly selected in each township, and 5–10 farmers were randomly surveyed from each village. Main questions in the survey questionnaire were given in two parts: (1) basic information of the orchard, including citrus yield, area, density, planting years, etc. (2) Orchard input, including fertilizer, pesticides, irrigation, the energy consumption of land preparation and weeding mechanical. Finally, a total of 155 valid questionnaires were collected.

2.2.2. Life Cycle Assessment

Life cycle assessment (LCA) is a method for analyzing the environmental impact of products from the cradle to the grave in their life cycle [29]. According to the ISO 14040, each LCA project has four essential phases: goal and scope definition, inventory analysis (inputs and outputs), impact assessment, and interpretation.

Goal, Scope Definition and Inventory Analysis

The study goal was assessing the life cycle of citrus production in Danling County, Sichuan, in southwest China. As for the system boundary, a cradle-to-farm-gate boundary was defined, which included two stages: the agricultural materials production stage (MS) and the arable farming stage (FS) (Figure 1). The MS included the production stages for fertilizers, pesticides, fuels and electricity, as well as transportation to the farm. The FS included the application of fertilizers (organic fertilizer and commercial synthetic fertilizer), pesticides, and the use of diesel by machinery. The functional units were per hectare and per ton of fresh citrus production.

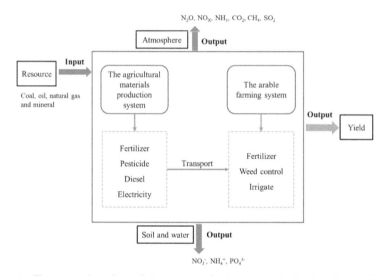

Figure 1. The system boundary of the citrus production system in Danling County, Sichuan, southwest China.

Impact Assessment

In this study, the global warming potential (GWP), acidification potential (AP), eutrophication potential (EP) were selected as evaluation objects, which were mainly driven by nutrient management [26]. The various life cycle impacts were calculated in accordance with International Organization for Standardization standards 14,040 and 14,044 [30,31].

The assessment formulas are as follows:

$$EI_t = EI_t \text{ direct} + EI_t \text{ indirect}, \tag{1}$$

where EI_t represents the total potentials for t environmental impact (EI) category, t (= 1, 2, 3) representing the impact category that includes GWP (kg CO_2-eq unit^{-1}), AP (kg SO_2-eq unit^{-1}) and EP (kg PO_4-eq unit^{-1}). EI_t direct refers to the potential emission value of environmental impact caused by inputs (fertilizers, pesticides, and diesel of machinery) in the process of their direct use in the arable farming stage. EI_t indirect means the potential emission value of environmental impact generated by various inputs in the production and transportation stages of agricultural materials.

$$EI_t \text{ direct} = \Sigma \, (IR_i \times EF_{iFS}), \tag{2}$$

$$EI_t \text{ indirect} = \Sigma \, (IR_i \times EF_{iMS}), \tag{3}$$

where IR_i represents the input rates (IR) of item i (fertilizers and pesticides consumption unit are kg, electricity consumption is kWh, diesel oil is L). EF_{iFS} means the emission factor (EF) of the input item i in the process of direct use in the arable farming stage. EF_{iMS} is the emission factors of the input of item i in the process of production and transportation. The emission factors of each input link on each environment are shown in Tables 1 and 2.

Table 1. Emission index of global warming potential, acidification potential, and eutrophication potential in the agricultural materials production stage.

Item	Unit	Global Warming (kg CO$_2$-eq Unit^{-1})	Acidification (kg SO$_2$-eq Unit^{-1})	Eutrophication (kg PO$_4$-eq Unit^{-1})	Reference [1]
Nitrogen production and transportation	kg N	8.28	0.0252	0.00303	[S1,S2]
Phosphorus production and transportation	kg P$_2$O$_5$	0.79	0.0006	0.00008	[S1,S2]
Potassium production and transportation	kg K$_2$O	0.55	0.00048	0.00006	[S1,S2]
Pesticides	kg	19.1	0.0105	0.00194	[S3,S4]
Diesel	L	3.75	0.0658	0.0119	[S4,S5]
Electricity	KW h	0.75	0.0145	0.00084	[S2,S6]

[1] Supplemental references for emission indexes are shown in the supplementary materials.

Table 2. The quantity of pollutants emitted (expressed as percentage of inputs) in the arable farming stage.

Pollution Emission	Emission Factors	References [1]
NH$_3$ emission	11.1% of nitrogen (N) fertilizer input	[S7]
NO$_3$ emission	9.97% of N fertilizer input	[S8]
N$_2$O emission		
Direct N$_2$O emission	1.25% of N fertilizer input	[S9,S10]
Indirect N$_2$O emission	1% NH$_3$ emission +2.5% NO$_3$ emission	[S10,S11]
NO$_X$ emission	10% of the N$_2$O emission	[S10]
Phosphorus loss	0.2% of total P$_2$O$_5$ fertilizer input	[S12,S13]

[1] Supplemental references for emission indexes are shown in the supplementary materials.

Result Interpretation

The excel 2016 and Duncan test by the Kruskal–Wallis one-way ANOVA in SPSS (20.0 version) were used to analysis the result of LCA. Results are explained at this phase to identify the key factors of the environmental impacts of citrus production in the Danling County, Sichuan, southwest China.

2.2.3. Farmer Grouping by Yield and Nitrogen Fertilizer Use Efficiency

The study adopted a farmer grouping method by yield and partial factor productivity of N fertilizer (PFP-N) [28] to analyze the relationship between various environmental impacts and related management practices among farmers. Based on average yield and PFP-N (Figure 2), the survey data of 155 farmers were divided into the following four groups: HH (high yield and high PFP-N), HL (high yield and low PFP-N), LH (low yield and high PFP-N), LL (low yield and low PFP-N) (Figure 1). The numbers of farmers in the LL, LH, HL, and HH groups were 62, 17, 30 and 46, respectively. The partial factor productivity of N fertilizer (PFP-N) (yield (kg ha^{-1}) divided by N application rate (kg ha^{-1})) means the amount of N fertilizer applied when producing unit yield, which is a significant index of N fertilizer efficiency in crop production [32,33].

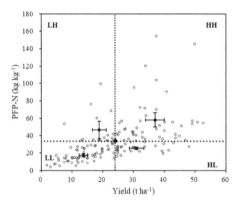

Figure 2. Relationships between citrus yield and partial factor productivity of nitrogen fertilizer (PFP-N) in Danling County, Sichuan, southwest China based on returned survey questionnaires from 155 farmers in 2017 and 2018. The black dotted lines represented the mean of citrus yield and PFP-N. Based on the average yield and PFP-N, the survey data obtained from the 155 farmers were divided into the following four groups: HH (high yield and high PFP-N), LH (low yield and high PFP-N), HL (high yield and low PFP-N), and LL (low yield and low PFP-N). The numbers of farmers in the LL, LH, HL, and HH groups were 62, 17, 30 and 46, respectively. The black circle represented the mean (with 95% confidence interval) of the yield and PFP-N in each group.

2.3. Experimental Design and Management

During the 2018 and 2019 cropping seasons, five field experiments were conducted in the main citrus producing area of Danling County, and two treatments were set in each field experiment: local fertilization recommendation (LR) and farmer practice (FP) in HH group. The soil properties used for field experiments were shown in Table S1. The fertilizer rate of LR treatment was determined basing on soil nutrient status and fertilizer recommendation by agricultural extension personnel of local government (Table S2). All field experiments were conducted in farmers' orchards. Except for fertilizer application, the treatments were managed by farmers in the same manner.

2.4. Data Analysis and Statistics

In order to compare the spatial distribution difference of environmental impacts in citrus orchards in Danling County, the spatial interpolation method of ArcGIS10.2 was used for geostatistical analyses (Figure 5). The excel 2016 was used to compare the data analysis and statistics. The Duncan test by the Kruskal–Wallis one-way ANOVA in SPSS (20.0 version) was used to test the significant differences at $p < 0.05$ (Figures 6, 7 and 10).

3. Results

3.1. The Input, Output and Environmental Impacts of Citrus Production System in Danling County

The investigated inputs and outputs of the citrus production in Danling County based on the 155 farmers' surveys are given in Table 3. For the input items, the average amounts (range) of N, phosphate (P), and potassium (K) fertilizer were 847 kg N ha^{-1} (140–2094 kg N ha^{-1}), 443 kg P$_2$O$_5$ ha^{-1} (84.6–1400 kg P$_2$O$_5$ ha^{-1}), and 693 kg K$_2$O ha^{-1}(130–1754 kg K$_2$O ha^{-1}), respectively. Chemical fertilizer accounted for 67.8–75.2% of total fertilizer input. In addition, the average input of pesticide, electricity used for pesticide application and irrigation, and diesel used for mechanical weeding was 21.2 kg ha^{-1}, 79.1 kWh ha^{-1} and 28.9 L ha^{-1}, respectively. For the output of citrus production (Table 3), the average yield (range) was 24.4 t ha^{-1} (1.88–56.3 t ha^{-1}), resulting in an averaged PFP-N of 34.0 kg kg^{-1}.

Table 3. Inputs and outputs of the citrus production in Danling County, Sichuan, southwest China based on returned survey questionnaires from 155 farmers in 2017 and 2018.

Item	Mean	Median	Range Max	Range Min	Standard Error
Input					
Total fertilizer (kg ha^{-1})					
N	847	802	2094	140	30.4
P_2O_5	443	395	1400	84.6	17.9
K_2O	693	643	1754	130	24.6
Chemical fertilizer (kg ha^{-1})					
N	598	603	1420	39.9	19.5
P_2O_5	324	286	953	0.00	14.0
K_2O	535	513	1237	0.00	18.7
Organic fertilizer (kg ha^{-1})					
N	249	169	1171	0.00	20.7
P_2O_5	119	75	846	0.00	11.2
K_2O	158	102	1171	0.00	14.8
Pesticide (kg ha^{-1})	21.2	17.8	96.6	0.43	1.36
Electricity (kWh ha^{-1})	79.1	57.0	327	5.42	5.48
Diesel (L ha^{-1})	28.9	18.8	169	0.00	2.45
Output					
Yield (t ha^{-1})	24.4	23.8	56.3	1.88	0.97
PFP-N (kg kg^{-1})	34.0	28.0	154	3.93	1.95

The invisible environmental cost of citrus production is shown in Table 4. The average GWP, AP, and EP were 11,665 kg CO_2-eq ha^{-1}, 184 kg SO_2-eq ha^{-1}, and 110 kg PO_4-eq ha^{-1}, respectively. For producing one ton of citrus fruit, the average values of GWP, AP, and EP were 642 kg CO_2-eq t^{-1}, 9.97 kg SO_2-eq t^{-1}, and 5.97 kg PO_4-eq t^{-1}, respectively. Each item of environmental impact showed great variability among farmers, as indicated by their range and standard error (Table 4).

Table 4. Main environmental impact (indicated by global warming, acidification, and eutrophication potential) expressed as per hectare or per ton of citrus production in Danling County, Sichuan, southwest China based on returned survey questionnaires from 155 farmers in 2017 and 2018.

Item	Mean	Median	Range Max	Range Min	Standard Error
Per hectare of the citrus production					
Global warming potential (kg CO_2-eq ha^{-1})	11,665	11,785	26,987	2422	349
Acidification potential (kg SO_2-eq ha^{-1})	184	176	445	32.6	6.26
Eutrophication potential (kg PO_4-eq ha^{-1})	110	105	271	18.8	3.90
Per ton of the citrus production					
Global warming potential (kg CO_2-eq t^{-1})	642	483	3629	90.9	41.2
Acidification potential (kg SO_2-eq t^{-1})	9.97	7.85	55.0	1.38	0.64
Eutrophication potential (kg PO_4-eq t^{-1})	5.97	4.67	33.1	0.84	0.38

Further analyses showed that the production and application of fertilizers contributed 45.6–46.2% and 46.2–49.5% to GWP, respectively. Similarly, the application of fertilizers was the major contributor to the AP (89.4–89.8 %) and EP (97.8–97.9 %) (Figure 3). Specific to the contribution of fertilizers to environmental costs, the production and application of N fertilizer accounted for more than 95% of the total environmental costs expressed either per hectare of citrus cropping area or per ton of yield production (Figure 4). In contrast, the inputs of pesticide, diesel, and electricity contributed less than 5% to GWP, AP, and EP under both function units (Figure 3).

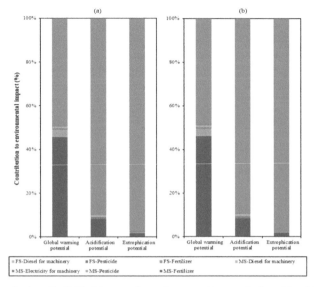

Figure 3. Contribution of individual input to the environmental impact (indicated by global warming, acidification, and eutrophication potential) in Danling County, Sichuan, southwest China based on returned survey questionnaires from 155 farmers when expressed as per hectare of citrus planted area (**a**) or per ton of citrus production (**b**). MS and FS represented the agricultural materials production stage and arable farming stage, respectively.

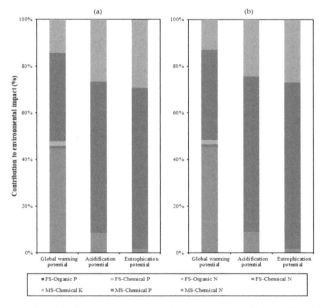

Figure 4. Contribution of fertilizers input to the environmental impact (indicated by global warming, acidification, and eutrophication potential) in Danling County, Sichuan, southwest China based on returned survey questionnaires from 155 farmers when expressed as per hectare of citrus planted area (**a**) or per ton of citrus production (**b**). MS and FS represented the agricultural materials production stage and arable farming stage, respectively.

Furthermore, the geographic information system (GIS) map showed large spatial variation in N fertilizer input and environmental costs (Figure 5). The spatial distribution of environmental costs of citrus production in Danling County was overall matched with that of N fertilizer input (Figure 5). In addition, the N input and environmental cost of citrus production in the newly developing areas of Danling County were lower than those in the major citrus production areas of Danling County (Figure 5; Supplementary Figure S1).

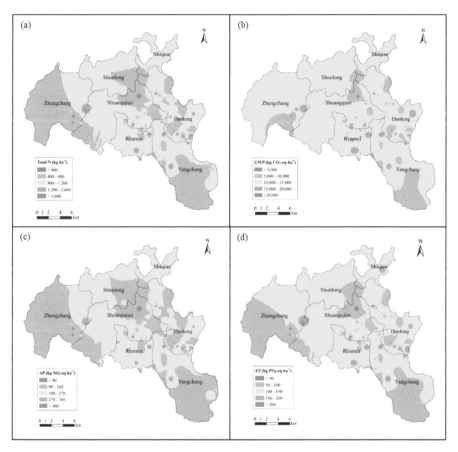

Figure 5. Geographical distribution of total nitrogen fertilizer input (**a**), global warming potential (**b**), acidification potential (**c**), and eutrophication potential (**d**) when expressed as per ton of citrus production in Danling County, Sichuan, southwest China based on returned survey questionnaires from 155 farmers in 2017 and 2018.

3.2. Potential of Emission Reduction Based on Grouping of Farmers Practice

Grouping methods according to citrus yield and PFP-N (Figure 2), the inputs and outputs of these four groups are shown in Table 5. The HH group had highest averaged yield (37.1 t ha^{-1}) and PFP-N (58.0 kg kg^{-1}), which were 52% and 70.6% higher than these average values of all 155 orchards, respectively. On the other hand, the HH group used fewer fertilizers than the other 155 orchards (Tables 3 and 5).

Table 5. Inputs and outputs of citrus production system among the four groups in Danling County, Sichuan, southwest China based on returned survey questionnaires from 155 farmers in 2017 and 2018.

Item	Orchard Group [1]			
	LL	LH	HL	HH
Input				
Total fertilizer (kg ha^{-1})				
N	857 ± 45.6b[2]	456 ± 41.5c	1247 ± 56.1a	717 ± 37.1b
P$_2$O$_5$	442 ± 28.9b	243 ± 31.0c	592 ± 34.5a	421 ± 29.6b
K$_2$O	707 ± 37.4b	427 ± 46.1c	923 ± 58.8a	624 ± 35.1b
Chemical fertilizer (kg ha^{-1})				
N	631 ± 31.3b	385 ± 41.6c	763 ± 31.4a	526 ± 32.5b
P$_2$O$_5$	323 ± 21.1a	203 ± 31.8b	389 ± 30.6a	327 ± 27.1a
K$_2$O	565 ± 30.6a	381 ± 46.1b	613 ± 38.7a	502 ± 32.7a
Organic fertilizer (kg ha^{-1})				
N	226 ± 25.7b	70.5 ± 16.8c	484 ± 64.0a	191 ± 28.5b
P$_2$O$_5$	119 ± 16.7b	39.9 ± 10.0c	203 ± 26.6a	94.4 ± 21.3bc
K$_2$O	142 ± 17.3b	45.6 ± 12.3c	310 ± 47.2a	122 ± 23.4bc
Pesticide (kg ha^{-1})	17.6 ± 1.84bc	15.7 ± 3.72c	26.7 ± 3.72a	24.4 ± 2.49ab
Electricity (kWh ha^{-1})	62.2 ± 7.91b	70.9 ± 14.9ab	99.0 ± 14.6a	92.1 ± 9.76ab
Diesel (L ha^{-1})	34.5 ± 4.71ab	44.1 ± 7.36a	20.0 ± 3.70b	21.6 ± 3.24b
Output				
Yield (t ha^{-1})	13.6 ± 0.71d	18.9 ± 1.07c	31.1 ± 1.19b	37.1 ± 1.51a
PFP-N (kg kg^{-1})	17.2 ± 0.93c	46.6 ± 4.61b	25.5 ± 0.83c	58.0 ± 4.07a

[1] Based on average yield and average partial factor productivity of nitrogen-fertilizer (PFP-N), the survey data obtained from the 155 farmers were divided into the following four orchard groups: HH (high yield and high PFP-N), LH (low yield and high PFP-N), HL (high yield and low PFP-N), and LL (low yield and low PFP-N). The numbers of orchards in the LL, LH, HL, and HH groups were 62, 17, 30 and 46, respectively. The PFP-N was an accessible index of N use efficiency in crop production, which was calculated as yield (kg ha^{-1}) divided by N application rate (kg ha^{-1}). [2] Values are means ± standard error and different letters after the values in the same row indicate significant differences between orchard groups at $p < 0.05$.

There were also significant differences in GWP, AP and EP among the four groups under both function units (Table 3, Table 5). When expressed as per ha of citrus cropping area, the average GWP of HH group was decreased by 14.1%, 34.8%, and 12.0% than that of LL group, HL group, and the average values of all 155 orchards, respectively. Similarly, the average AP was decreased by 16.1%, 41.0%, and 14.8%, respectively; and the EP was decreased by 16.2%, 42.0%, and 15.2%, respectively (Figure 6). When expressed as per ton of yield production, the average GWP of HH group was decreased by 72.5%, 25.4%, 44.9% and 55.5% than that of LL group, LH group, HL group, and the average values of all 155 orchards, respectively. Similarly, the average AP was decreased by 73.0%, 20.3%, 49.9%, and 56.6%, respectively; and the EP was decreased by 73.0%, 18.1%, 50.7%, and 56.6%, respectively (Figure 7).

Less application of N fertilizer, especially chemical N fertilizer resulted in lower environmental costs in HH group (Figures 6 and 7). Such management was related with among HH group orchards and other orchards (Figure 8). In term of farmer age, the proportion of farmers less than 50 years old was 58% in HH group, which was higher than that in other groups. For education level, 84% of family major labor in HH group received junior high school and above education, which was better than other groups. In addition, more than 64% of orchards are managed by less than two labors. For orchard conditions, high yield orchards were planted in sloping and/or flat fields with proper planting density (625–1112 plant ha^{-1}). Furthermore, the proportion of citrus orchards fertilized less than four times was 47% in HH group, which was obviously higher than other groups (Figure 7).

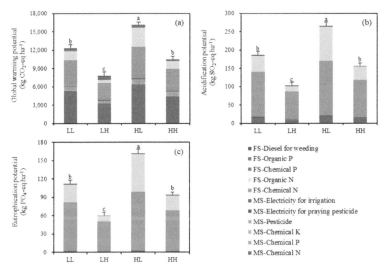

Figure 6. Global warming potential (**a**), acidification potential (**b**), and eutrophication potential (**c**) per hectare of citrus production in four groups of investigated farmers in Danling County, Sichuan, southwest China. The four groups included: HH (high yield and high PFP-N), LH (low yield and high PFP-N), HL (high yield and low PFP-N), and LL (low yield and low PFP-N) groups, respectively. MS and FS represented the agricultural materials production stage and arable farming stage, respectively. The bars were means + standard error; The column with different letters indicated significant difference at $p < 0.05$.

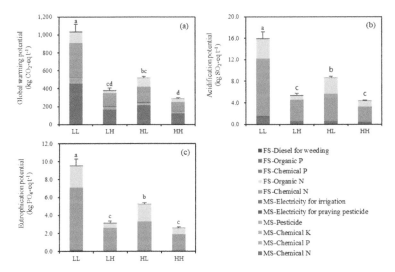

Figure 7. Global warming potential (**a**), acidification potential (**b**), and eutrophication potential (**c**) per ton of citrus production in four groups of investigated farmers in Danling County, Sichuan, southwest China. The four groups included: HH (high yield and high PFP-N), LH (low yield and high PFP-N), HL (high yield and low PFP-N), and LL (low yield and low PFP-N) groups, respectively. MS and FS represented the agricultural materials production stage and arable farming stage, respectively. The bar indicated mean + standard error; The columns with different letters indicated significant difference at $p < 0.05$.

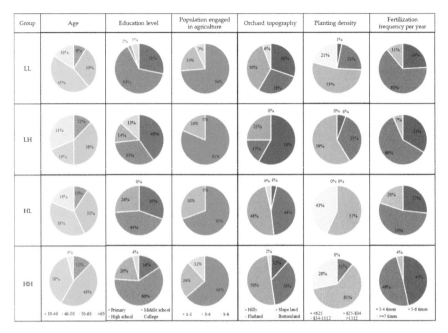

Figure 8. Farmer practice for citrus production among the four groups in Danling County, Sichuan, southwest China based on returned survey questionnaires from 155 farmers in 2017 and 2018. The investigated items of farmer practice included age, education level, population engaged in agriculture, orchard topography, planting density (plants ha^{-1}), and fertilization frequency per year. The four groups included: HH (high yield and high PFP-N), LH (low yield and high PFP-N), HL (high yield and low PFP-N), and LL (low yield and low PFP-N) groups, respectively.

3.3. The Environmental Impacts of the Citrus Production with Local Recommendation of Fertilization

Five field demonstrations with local recommendation of fertilization (LR) were further conducted to explore the potential to reduce environmental impact based on orchards of HH group. Under the same farmer management practice except fertilization, the citrus yield with LR treatment was increased by 1.9–49.5% than that with farmers' fertilization practice, and the PFP-N with LR treatment was increased by 38.0–116% than that with the farmers' fertilization practice (Figure 9). Whereas, the fertilizer input of LR treatment was decreased by 13.4–30.8% than that of the farmers' fertilization practice (Supplementary Table S2).

Taken together, the LR treatment resulted in significantly lower environmental impacts than the farmer's fertilization practice (Figure 10). When expressed as per ha of citrus cropping area, comparing with the FP treatment, the average values of GWP, AP and EP in the LR treatment were decreased by 21.2%, 23.1%, and 23.5%, respectively. When expressed as per ton of yield production, comparing with the FP, the average values of GWP, AP and EP with the LR treatment were decreased by 33.5%, 34.8%, and 35.2%, respectively.

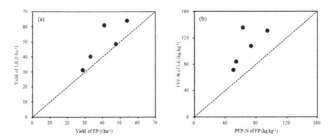

Figure 9. The correlation of yield (**a**) and PFP-N (**b**) between the local recommendation (LR) treatment and farmer practice (FP) treatment in five field experiments in Danling County, Sichuan, southwest China. The PFP-N was an index of nitrogen (N) use efficiency in crop production, which was calculated as yield (kg ha^{-1}) divided by N application rate (kg ha^{-1}).

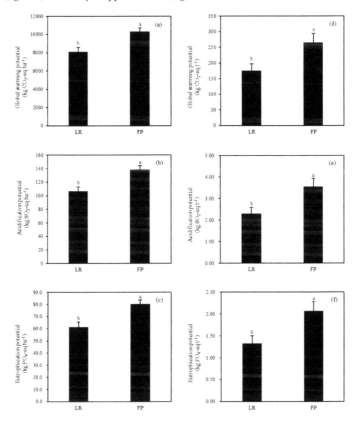

Figure 10. The means of global warming potential (**a, d**), eutrophication potential (**b, e**), and acidification potential (**c, f**) in five field experiments when expressed as per hectare of citrus production (**a–c**) or per ton of the citrus production (**d–f**). LR and FP represented treatments of local recommendation and farmer practice, respectively. The bar indicated mean + standard error; The columns with different letters indicated significant difference at $p < 0.05$.

4. Discussion

4.1. A High Environmental Risk Existed in Citrus Production System in Danling County

The findings of this study indicate a high environmental risk for citrus production in Danling County (Table 4). Under the same system boundary in China, the value (expressed per ha of citrus cropping area) of the GWP of citrus production in Danling County was about 1.2–2.0 times higher than that of other fruit production [22,27], and about 2.9–5.1 times higher than that of vegetable and cereal crop production [22,25,26,34,35]. Specific to the citrus production system, the value (expressed per hectare of citrus cropping area) of the GWP was 1.6, 1.5, and 1.1 times higher than the previous reports of citrus in China [22], Iran [21], and Spain [21,24]. Similarly, higher values of AP and EP were also found in citrus production in Danling County when compared with previous studies conducted in other countries, including Iran [21], Spain [23,24,36], and Italy [23]. Furthermore, the environmental costs (expressed per one ton of citrus yield production) in this study were even higher than previous studies [21,24,26,27,37]. Considering the substantial difference of environmental cost in citrus production in Danling County and other regions [22], the whole environmental cost of the citrus production in China deserves further studies, with a detection of more hotspots at the county level.

The production and utilization of fertilizer were the most important factor to the environmental impacts which contributed 92.4–95.1%, 89.4–89.8% and 97.8–97.9% to the potential of GWP, AP and EP, respectively (Figure 3). This was consistent with previous studies finding that fertilizer related processes were the major contribution to environmental costs [21,22,26,27]. However, the extremely high fertilizer input resulted in substantially higher contribution to the environmental costs (Tables 3 and 4). The average inputs of chemical N, P_2O_5, and K_2O fertilizer for citrus production were about 1.5–3.6, 1.33–6.3, and 4.0–5.2 times higher in this study than that in previous studies in other countries [21,24], and also higher than those for horticultural crops and cereal crops in China [22,25,26,34]. The reason for the high fertilizer input was due to the high benefit of citrus production and their misconception about plant requirements for nutrients. It is generally believed by farmers that a greater fertilizer input would yield more products and greater returns [4,38]. Furthermore, the production and application of N fertilizer accounted for more than 95% of the total costs derived from fertilizer related processes (Figures 3 and 4), thus the hotspots of environmental risk were spatially correlated with N fertilizer input (Figure 5). Therefore, the first option for reducing environmental impacts of the citrus production in this county would be to optimize nutrient management and reduce the fertilizer input.

4.2. Great Potential of Reducing Environmental Impact by Learning from Good Farmers

The great variation in citrus yield, PFP-N, environmental costs of citrus production within Danling County (Tables 3 and 4; Figures 2 and 5) raised the possibility to reduce the environmental impact by learning from excellent farmers [26,27,39]. The HH group had better performances in citrus yield, PFP-N, but with less fertilizer input and consequential lower environmental costs (Table 5; Figures 6 and 7). These differences might be explained by their management. On average, farmers in HH group are younger and have received a better education than farmers in other groups (Figure 8), and are conducive to the faster acquisition, update and apply new knowledge and technology [4]. In addition, the farmers in HH group applied fertilizer with less application times than others (Figure 8), which was correlated with less fertilizer input (Table 5). For the orchard establishment, orchards planted in sloping and/or flat fields with proper planting density were easier to realize high yield and high efficiency of nutrient management (Figure 8; Table 5). Therefore, learning the management pattern of HH group by other groups would be useful for formulation of regional industrial policies and demonstration to simultaneously achieve high yield, high economic benefit and low environmental cost at the county level.

4.3. Further Potential to Reduce Environmental Impact by Local Recommendation of Fertilization

It is noticeable that there is substantial potential to further reduce environmental impact of the citrus production from the HH group (Table 5), because such an impact was still much higher than previous studies [21,24]. Preliminary demonstrations conducted by local government and Danling STB have confirmed the possibility. With less fertilizer input (Table S1), the LR treatment resulted in higher citrus yield, higher PFP-N, and consequently lower environmental impacts than the farmer's fertilization practice (Figure 9, Figure 10). Such results are consistent with previous studies indicating that cereal production with high yield and low environmental cost could be realized by integrative soil-crop nutrient management [25,39]. Considering that the inputs of N, P, and K fertilizer (Table S1) in the LR treatment is still higher than that in the expert recommendation in China [4], the 13–24% reduction of fertilizer input from farmer practices to the LR treatment was expected to have no negative effect on yield. A previous study also revealed that there were large opportunities to reduce the environmental impact of agriculture by eliminating nutrient overuse, while allowing for an approximately 30% increase in the production of major cereals [40]. Thus, it is reasonable to predict a further reduction of environmental cost for citrus production in Danling County, by optimizing fertilization strategy, including the right rate, right source, right place, and right time [41]. For example, citrus yield, N use efficiency, and water use efficiency can be greatly improved by fertigation technology [42].

4.4. The Last Mile to Realize the Sustainable Citrus Production in China's County Level

The adoption of advanced knowledge and management techniques by farmers is the vital key to achieve sustainable citrus production at the county level in China. This study has provided evidence in facilitating the role of technical training in the adoption of more environmentally-friendly practices in agriculture [43]. Enhanced management measures had potential impacts on the sustainable development of agriculture under the smallholder farming practice [39,44]. Thus, the Danling STB was set up in 2017 to explore the limiting factors of the sustainable citrus production, and to conduct field trials, technical training, field demonstration and participatory innovation (Supplementary Figure S1). The farmers are intended to adopt recommended management practices when these limitations and farmers' concerns are addressed [45]. In addition, policymakers should realize appropriate policy interventions and adjustments, such as enacting regulatory policies related to the use of fertilizer [42]. Finally, new media like Wechat, mobile Apps, and short videos could also be used to publicize the specific measures of scientific agronomic practices. Through these measures, the last mile could be eliminated to be close to sustainable citrus production in China's county level.

5. Conclusions

This study was the first attempt to address the hotspots and optimization of the environmental cost of a typical citrus-producing county in China. The findings indicated that high environmental risks existed in the citrus production in the studied Danling County. Fertilization was identified as the most important factor contributing to the GWP, AP and EP under both function units. Meanwhile, the spatial distribution of environmental cost in this county was well matched with that of N input. Under the scenario of learning from excellent farmers, the HH orchard group with younger and more educated owners achieved better performances in citrus yield, PFP-N, while with less fertilizer input and lower environmental cost. Preliminary field experiments had confirmed that optimizing nutrient management could achieve higher yield, higher nutrient use efficiency, and lower environmental cost than the traditional fertilization practice in HH group at the local level. As the connex of local government, university, agro-material supplier and farmers, the Danling STB might facilitate to eliminate the current obstacles to realize the sustainable citrus production at county level. Overall, this study highlighted the importance to optimize farmer practices, especially their nutrient management in citrus production, which should be crucial for achieving the win-win target of productivity and environmental sustainability. Considering the substantial differences in the citrus production, including soil type, climate condition, and orchard management in different regions, the whole environmental cost of the citrus production in China deserves further studies with more hotspots at the county level.

Supplementary Materials: The following information is available online at http://www.mdpi.com/2071-1050/12/5/1827/s1, Figure S1, the locations of the citrus production in Danling County and the Danling Science and Technology backyards (Danling STB); Table S1, the soil properties used for field experiments at five sites; Table S2, the fertilizer inputs of five field experiments; and Supplemental references for emission index.

Author Contributions: M.Y., Q.L., Z.W., G.Z. and Z.W. conducted the farmers survey with the assistance by Y.Z., X.S., B.L. and J.Q.; M.Y. and Q.L. conducted the field experiments; J.W., C.G. and X.W. provided the methodology of Life Cycle Assessment and the emission factors and calculation formula template; M.Y. analyzed the data and wrote the manuscript with the assistance by Q.L., W.L. and H.X.; X.C., X.S., F.Z. and Y.Z. designed, supervised this study; X.H. revised and approved this manuscript for publication; all authors have read and agreed to the published version of the manuscript. All authors have read and agreed to the published version of the manuscript.

Funding: This study was supported by the National Key Research and Development Project of China (2016YFD0200104), the National Key Technology Research and Development Program (2015BAD06B04), the Fundamental Research Funds for the Central Universities (2362015xk06, XDJK2013C065 and 20710922).

Acknowledgments: The authors also acknowledge with gratitude Danling County Agricultural and Rural Bureau in data collection in Danling County, Qingsong Zhang and Zhanqing Zhao in preparation of the GIS maps, and Wei Qin for his advice in writing this paper.

Conflicts of Interest: The authors declare no conflict of interest.

References

1. Food and Agriculture Organization of the United Nations. *FAO Statistical Yearbook 2013*; World Food and Agriculture; Food and Agriculture Organization of the United Nations: Rome, Italy, 2017.
2. Shen, Z.M. Industry status and demonstration leading role of top 30 citrus counties in China. *Fruit Grow. Friend* **2019**, *3*, 1–4.
3. China Agriculture Statistical Report 2018. Available online: http://www.stats.gov.cn/tjsj/ndsj/2019/indexch.htm (accessed on 20 December 2019).
4. Li, Y.J.; Yang, M.; Zhang, Z.Z.; Li, W.L.; Guo, C.Y.; Chen, X.P.; Shi, X.J.; Zhou, P.; Tang, X.D.; Zhang, Y.Q. An Ecological Research on Potential for Zero-growth of Chemical Fertilizer Use in Citrus Production in China. *Ekoloji* **2019**, *28*, 1049–1059.
5. Lei, J.; Liang, S.S.; Tan, Q.L.; Hu, C.X.; Sun, X.C.; Zhao, X.H. NPK fertilization rates and reducing potential in the main citrus producing regions of China. *J. Plant Nutr. Fertil.* **2019**, *25*, 1504–1513.
6. Mattos, J.D.; Quaggio, J.A.; Cantarella, H. Nutrient management for high citrus fruit yield in tropical soils. *Better Crop. Plant Food* **2012**, *96*, 4–7.
7. Obreza, T.A.; Morgan, K.T. *Nutrition of Florida Citrus Trees*, 2nd ed.; University of Florida: Gainesville, FL, USA, 2008.
8. Tilman, D.; Cassman, K.G.; Matson, P.A.; Naylor, R.; Polasky, S. Agricultural sustainability and intensive production practices. *Nature* **2002**, *418*, 671–677. [CrossRef] [PubMed]
9. Guo, J.H.; Liu, X.J.; Zhang, Y.; Shen, J.L.; Han, W.X.; Zhang, W.F.; Christie, P.; Goulding, K.W.T.; Vitousek, P.M.; Zhang, F.S. Significant acidification in major Chinese croplands. *Science* **2010**, *327*, 1008–1010. [CrossRef] [PubMed]
10. Huang, J.; Xu, C.C.; Ridoutt, B.G.; Wang, X.C.; Ren, P.A. Nitrogen and phosphorus losses and eutrophication potential associated with fertilizer application to cropland in China. *J. Clean. Prod.* **2017**, *159*, 171–179. [CrossRef]
11. Li, B.; Fan, C.H.; Zhang, H.; Chen, Z.Z.; Sun, L.Y.; Xiong, Z.Q. Combined effects of nitrogen fertilization and biochar on the net global warming potential, greenhouse gas intensity and net ecosystem economic budget in intensive vegetable agriculture in southeastern China. *Atmos. Environ.* **2015**, *100*, 10–19. [CrossRef]
12. Lv, Y.H.; Zhang, C.; Ma, J.N.; Yun, W.J.; Gao, L.L.; Li, P.S. Sustainability assessment of smallholder farmland systems: Healthy farmland system assessment framework. *Sustainability* **2019**, *11*, 4525. [CrossRef]
13. Youssef, A.M.; Abu Abdullah, M.M.; Pradhan, B.; Gaber, A.F.D. Agriculture sprawl assessment using multi-temporal remote sensing images and its environmental impact; Al-Jouf, KSA. *Sustainability* **2019**, *11*, 4177. [CrossRef]

14. Van Cauwenbergh, N.; Biala, K.; Bielders, C.; Brouckaert, V.; Franchois, L.; Cidad, V.G.; Hermy, M.; Mathijs, E.; Muys, B.; Reijnders, J.; et al. SAFE—A hierarchical framework for assessing the sustainability of agricultural systems. *Agric. Ecosyst. Environ.* **2007**, *120*, 229–242. [CrossRef]

15. Fan, W.G.; Zhang, P.; Xu, Z.H.; Wei, H.J.; Lu, N.C.; Wang, X.C.; Weng, B.Q.; Chen, Z.D.; Wu, F.L.; Dong, X.B. Life cycle environmental impact assessment of circular agriculture: A case study in Fuqing, China. *Sustainability* **2018**, *10*, 1810. [CrossRef]

16. Haas, G.; Wetterich, F.; Geier, U. Life Cycle Assessment Framework in Agriculture on the Farm Level. *Int. J. Life Cycle Assess.* **2000**, *5*, 345–348. [CrossRef]

17. Guinee, J.B.; Heijungs, R.; Huppes, G.; Zamagni, A.; Masoni, P.; Buonamici, R.; Ekvall, T.; Rydberg, T. Life Cycle Assessment: Past, Present, and Futures. *Environ. Sci. Technol.* **2011**, *45*, 90–96. [CrossRef]

18. Cerutti, A.K.; Beccaro, G.L.; Bruun, S.; Bosco, S.; Donno, D.; Notarnicola, B.; Bounous, G. Life cycle assessment application in the fruit sector: State of the art and recommendations for environmental declarations of fruit products. *J. Clean. Prod.* **2014**, *73*, 125–135. [CrossRef]

19. Bessou, C.; Basset-Mens, C.; Latunussa, C.; Velu, A.; Heitz, H.; Vanniere, H.; Caliman, J.P. Partial modelling of the perennial crop cycle misleads LCA results in two contrasted case studies. *Int. J. Life Cycle Assess.* **2016**, *21*, 297–310. [CrossRef]

20. Sanjuan, N.; Ubeda, L.; Clemente, G.; Mulet, A.; Girona, F. LCA of integrated orange production in the Comunidad Valenciana (Spain). *Int. J. Agric. Resour. Gov. Ecol.* **2005**, *4*, 163–177. [CrossRef]

21. Alishah, A.; Motevali, A.; Tabatabaeekoloor, R.; Hashemi, S.J. Multiyear life energy and life cycle assessment of orange production in Iran. *Environ. Sci. Pollut. Res.* **2019**, *26*, 32432–32445. [CrossRef]

22. Yan, M.; Cheng, K.; Yue, Q.; Yan, Y.; Rees, R.M.; Pan, G.X. Farm and product carbon footprints of China's fruit production-life cycle inventory of representative orchards of five major fruits. *Environ. Sci. Pollut. Res.* **2016**, *23*, 4681–4691. [CrossRef]

23. Nicolo, B.F.; De Salvo, M.C.; Ramirez-Sanz, C.; Estruch, V.; Sanjuan, N.; Falcone, G.; Strano, A. Life cycle assessment applied to different citrus farming systems in Spain and Italy. *Agroecol. Sustain. Food Syst.* **2018**, *42*, 1092–1105. [CrossRef]

24. Ribal, J.; Ramirez-Sanz, C.; Estruch, V.; Clemente, G.; Sanjuan, N. Organic versus conventional citrus. Impact assessment and variability analysis in the Comunitat Valenciana (Spain). *Int. J. Life Cycle Assess.* **2017**, *22*, 571–586. [CrossRef]

25. Chen, X.P.; Cui, Z.L.; Fan, M.S.; Vitousek, P.; Zhao, M.; Ma, W.Q.; Wang, Z.L.; Zhang, W.J.; Yan, X.Y.; Yang, J.C.; et al. Producing more grain with lower environmental costs. *Nature* **2014**, *514*, 486–489. [CrossRef] [PubMed]

26. Wang, X.Z.; Zou, C.Q.; Zhang, Y.Q.; Shi, X.J.; Liu, J.Z.; Fan, S.S.; Liu, Y.M.; Du, Y.F.; Zhao, Q.Y.; Tan, Y.G.; et al. Environmental impacts of pepper (*Capsicum annuum* L.) production affected by nutrient management: A case study in southwest China. *J. Clean. Prod.* **2018**, *171*, 934–943. [CrossRef]

27. Guo, C.Y.; Wang, X.Z.; Li, Y.J.; He, X.H.; Zhang, W.S.; Wang, J.; Shi, X.J.; Chen, X.P.; Zhang, Y.Q. Carbon footprint analyses and potential carbon emission reduction in China's major peach orchards. *Sustainability* **2018**, *10*, 2908. [CrossRef]

28. Ye, Y.L.; Wang, G.L.; Huang, Y.F.; Zhu, Y.J.; Meng, Q.F.; Chen, X.P.; Zhang, F.S.; Cui, Z.L. Understanding physiological processes associated with yield-trait relationships in modern wheat varieties. *Field Crop. Res.* **2011**, *124*, 316–322. [CrossRef]

29. Masuda, K. Eco-efficiency assessment of intensive rice production in japan: Joint application of life cycle assessment and data envelopment analysis. *Sustainability* **2019**, *11*, 5368. [CrossRef]

30. International Organization for Standardization (ISO). *Environmental Management-Life Cycle Assessment-Principles and Framework*; ISO 14040: 2006; Quality Press: Milwaukee, WI, USA, 2014.

31. International Organization for Standardization (ISO). *Environmental Management-Life Cycle Assessment-Requirements and Guidelines*; ISO 14044: 2006; Quality Press: Milwaukee, WI, USA, 2014.

32. Cui, Z.L.; Wang, G.L.; Yue, S.C.; Wu, L.; Zhang, W.F.; Zhang, F.S.; Chen, X.P. Closing the N-Use Efficiency Gap to Achieve Food and Environmental Security. *Environ. Sci. Technol.* **2014**, *48*, 5780–5787. [CrossRef]

33. Holka, M.; Jankowiak, J.; Bienkowski, J.F.; Dabrowicz, R. Life cycle assessment (LCA) of winter wheat in an intensive crop production system in Wielkopolska region (Poland). *Appl. Ecol. Environ. Res.* **2016**, *14*, 535–545. [CrossRef]

34. Yan, M.; Cheng, K.; Luo, T.; Yan, Y.; Pan, G.X.; Rees, R.M. Carbon footprint of grain crop production in China—Based on farm survey data. *J. Clean. Prod.* **2015**, *104*, 130–138. [CrossRef]

35. Romero-Gamez, M.; Audsley, E.; Suarez-Rey, E.M. Life cycle assessment of cultivating lettuce and escarole in Spain. *J. Clean. Prod.* **2014**, *73*, 193–203. [CrossRef]
36. Beltran-Esteve, M.; Reig-Martinez, E.; Estruch-Guitart, V. Assessing eco-efficiency: A metafrontier directional distance function approach using life cycle analysis. *Environ. Impact Assess. Rev.* **2017**, *63*, 116–127. [CrossRef]
37. Cai, Y.J.; Qiao, Y.H.; Xu, J.; Meng, F.Q.; Wu, W.L. Environmental impact assessment via life cycle analysis for organic and conventional apple productions. *Chin. J. Eco-Agric.* **2017**, *25*, 1527–1534.
38. Zhao, P.F.; Cao, G.X.; Zhao, Y.; Zhang, H.Y.; Chen, X.P.; Li, X.L.; Cui, Z.L. Training and organization programs increases maize yield and nitrogen-use efficiency in smallholder agriculture in China. *Agron. J.* **2016**, *108*, 1944–1950. [CrossRef]
39. Cui, Z.L.; Zhang, H.Y.; Chen, X.P.; Zhang, C.C.; Ma, W.Q.; Huang, C.D.; Zhang, W.F.; Mi, G.H.; Miao, Y.X.; Li, X.L.; et al. Pursuing sustainable productivity with millions of smallholder farmers. *Nature* **2018**, *555*, 363–366. [CrossRef] [PubMed]
40. Mueller, N.D.; Gerber, J.S.; Johnston, M.; Ray, D.K.; Ramankutty, N.; Foley, J.A. Closing yield gaps through nutrient and water management. *Nature* **2012**, *490*, 254–257. [CrossRef]
41. Xu, X.P.; He, P.; Pampolino, M.F.; Johnston, A.M.; Qiu, S.J.; Zhao, S.C.; Chuan, L.M.; Zhou, W. Fertilizer recommendation for maize in China based on yield response and agronomic efficiency. *Field Crop. Res.* **2014**, *157*, 27–34. [CrossRef]
42. Qin, W.; Assinck, F.B.T.; Heinen, M.; Oenema, O. Water and nitrogen use efficiencies in citrus production: A meta-analysis. *Agric. Ecosyst. Environ.* **2016**, *222*, 103–111. [CrossRef]
43. Liu, Y.; Ruiz-Menjivar, J.; Zhang, L.; Zhang, J.B.; Swisher, M.E. Technical training and rice farmers' adoption of low-carbon management practices: The case of soil testing and formulated fertilization technologies in Hubei, China. *J. Clean. Prod.* **2019**, *226*, 454–462. [CrossRef]
44. Oliver, D.M.; Zheng, Y.; Naylor, L.A.; Murtagh, M.; Waldron, S.; Peng, T. How does smallholder farming practice and environmental awareness vary across village communities in the karst terrain of southwest China? *Agric. Ecosyst. Environ.* **2020**, *288*, 106715. [CrossRef]
45. Zhang, W.F.; Cao, G.X.; Li, X.L.; Zhang, H.Y.; Wang, C.; Liu, Q.Q.; Chen, X.P.; Cui, Z.L.; Shen, J.B.; Jiang, R.F.; et al. Closing yield gaps in China by empowering smallholder farmers. *Nature* **2016**, *537*, 671–674. [CrossRef]

Article

Net Greenhouse Gas Emissions from Agriculture in China: Estimation, Spatial Correlation and Convergence

Haoyue Wu [1,*], Hanjiao Huang [2], Jin Tang [1], Wenkuan Chen [1,*] and Yanqiu He [1]

1   College of Management, Sichuan Agricultural University, Chengdu 611130, China
2   College of Forestry, Northwest A&F University, Xianyang 712100, China
*   Correspondence: tsuki710064315@163.com (H.W.); 11454@sicau.edu.cn (W.C.)

Received: 15 June 2019; Accepted: 30 August 2019; Published: 4 September 2019

Abstract: The agricultural ecosystem has dual attributes of greenhouse gas (GHG) emission and absorption, which both influence the net amount of GHG. To have a clearer understanding of the net GHG effect, we linked up the emission and absorption of the agricultural ecosystem, estimated the net emissions of 30 provinces in China from 2007 to 2016, then explored the spatial correlation from global and local perspectives by Moran's I, and finally tested the convergence of the net emissions by α convergence test, conditional β convergence test and spatial econometric methods. The results were: (1) The average of provincial agricultural net GHG emissions was around 4999.916×10^4 t, showing a fluctuating trend in the 10 years. Meanwhile, the gaps among provinces were gradually widening, as the provinces with high emissions were mainly agglomerated in the middle reaches of the Yangtze River, while those with less emissions mainly sat in the northwest. (2) The net emissions correlated spatially in close provinces. The agglomeration centers were located in the middle reaches of the Yangtze River and the northern coastal region, showing "high–high" and "low–low" agglomeration, respectively. (3) The net emissions did not achieve α convergence or conditional β convergence in the whole country, but the growth rate had a significant positive spillover effect among adjacent provinces, and two factors, the quantity of the labor force and the level of agricultural economy, had a negative impact on the rate. It is suggested that all provinces could strengthen regional cooperation to reduce agricultural net GHG emissions.

Keywords: net greenhouse gas emissions; agriculture; spatial correlation; Moran's I; α convergence; conditional β convergence

1. Introduction

Global warming has become the most severe environmental issue of the world. As greenhouse gas (GHG) is the main factor for temperatures rising, countries around the world have realized the urgency of low-carbon development. As the largest emitter of GHG, China aims to reduce 60%~65% GHG emissions per unit of GDP by 2030 compared to 2005. Emission mitigation should start with its sources. Industry is recognized as the main source of GHG emissions, while agriculture also releases significant amounts of CO_2, CH_4 and N_2O to the atmosphere [1]. GHG emissions from agriculture in China, accounting for 17% of the total amount [2] and predicted to increase by another 30% in 2050 [3], should not be neglected. Meanwhile, the agricultural ecosystem also plays an important role in absorbing GHG [4], which influences the total amount, too. The complicated process of agricultural production makes it impossible to judge the net effect of GHG directly. In addition, because of the vast territory of China, natural resources and agricultural development may be completely different among regions, broadening the gaps of agricultural net GHG amount, but the geographical environment and industrial structure of neighboring provinces are usually similar, which may cause the net amount to correlate spatially. Therefore, what is the spatial–temporal characteristic of agricultural net GHG amount in China? Will it show spatial correlation? How is it evolving? Perhaps which regional emission mitigation measures to take can be addressed by answering the above questions.

To clarify the sources and to estimate the amount of GHG from agriculture, scholars have carried out relevant research. They have agreed that agricultural GHG emission comes from the process of agricultural production, mainly involving crop and livestock production, forestry and farmland use [3], also including agricultural waste, agricultural energy use and bio-burning [1,4]. There are two major perspectives for estimating: one is to focus on sources of a certain category, such as the cropping system [5–7], farmland use [8], agricultural burning [9], waste products [10] and livestock [11–13]. The other is to measure the total amount of a variety of sources from agriculture, which takes a certain country or region as the research object [14–16]. As for the GHG-absorption function of the agro-ecosystem, scholars generally believe that the absorption is contributed by forest, grassland and farmland. Research of Europe and the United States focus on forest, including soil carbon sequestration [17,18] and forest carbon sinks [19–21]. In China, scholars mainly consider absorption of cropping system, that is, the GHG absorption of crops by photosynthesis during the growth cycle [22,23]. In general, related studies are consistent in the coefficients of crops' absorption [4]. Based on the estimation of GHG emission and absorption, scholars have begun to link them up and explore the net effect in different countries or regions [24–28]. In the representative research of China, Duan et al. study the carbon emissions, absorption and footprint of farmland ecosystems, and find that they all show an increasing trend [29]. Chen et al. measure the carbon absorption and emission of agricultural systems from 1991 to 2011, deeming the system is a sink of GHG [30]. The existing studies have laid a foundation for thorough exploration on the net GHG emissions from agriculture. However, the consideration of sources and estimation methods have not been unified, resulting in different conclusions.

After understanding the sources and amount of agricultural GHG emission, scholars have paid attention to its regional disparities [31,32], finding several factors that lead to the regional difference, such as technical progress [33], production mode [34], agricultural practitioners, disaster degree, industrial structure, economic development and public investment [35], and then proposing a series of potential mitigation options [36,37]. With the research deepening, the convergence test, a tool to investigate the evolution of regional disparities, is applied to analyze the regional difference of GHG emissions [38–40]. Yang estimates the amount and intensity of agricultural carbon emissions in China from 1993 to 2011, and confirms that there is no α convergence and conditional β convergence in the intensity [41]. Cheng et al. study the convergence trend of agricultural carbon productivity in China from 1997 to 2012, whose results show that there is no α convergence but absolute β convergence [42]. While Wu et al. take the slack based measure under undesirable outputs (SBM-Undesirable) to evaluate China's agricultural carbon emissions' performance from 2000 to 2014 and believe that there is no stochastic convergence in carbon emissions or its performance [43]. As seen from the convergence study, scholars concentrate on agricultural GHG emission, paying insufficient attention to absorption, and seldom consider the net effect. Besides, scholars assume that emissions are independent in different regions, so they apply ordinary panel econometric methods when conducting the β convergence test. In fact, agricultural GHG is more likely to correlate spatially because of the similar resource endowments, industrial structure and the emission-mitigation policy imitation in neighboring provinces. If the potential spatial correlation is ignored, it may affect the accuracy of the results [44].

To make up for the existing research, we comprehensively took 21 sources of agricultural GHG emissions into consideration, and linked up the emission and absorption of agricultural ecosystem, to estimate the net GHG emissions of 30 provinces in China from 2007 to 2016. Then, we chose the Moran's I based on the distance reciprocal square matrix to explore the spatial agglomeration of the emissions. Finally, the convergence theory and spatial econometric methods were used to analyze the emissions' convergence, aiming to offer a reference for controlling the emissions from both temporal and spatial perspectives. The article is structured as: Section 2 introduces the method and data involved in this study. The third section presents the empirical analysis. The fourth section discusses the results and the last section gives the conclusions.

2. Method and Data

2.1. Calculation of Agricultural Net GHG Emission

Agricultural GHG emission refers to the greenhouse gases released into atmosphere by agricultural production, while agricultural GHG absorption is the greenhouse gases absorbed by crops through photosynthesis, especially carbon dioxide. Both of them act on the net amount, when GHG released minus that of absorbed is positive, agriculture ecosystem is a GHG source, and the corresponding amount is called net emission. On the contrary, when the agriculture ecosystem becomes a GHG sink, the corresponding amount is net absorption.

2.1.1. Agricultural GHG Emission

Four major categories of agricultural GHG emission sources were considered, totaling 21 items. First, carbon emission caused by farmland utilization, involving the input of agricultural materials, the plowing and irrigation activities; second, CH_4 produced during the growth of rice; third, CH_4 and N_2O from intestinal fermentation and fecal management in ruminant breeding; fourth, carbon emissions from agricultural energy. The total GHG emission was the sum of the emissions from 21 sources of four categories mentioned above.

The emissions of each category can be calculated as:

$$E_i = \sum E_j = \sum (T_j \cdot \delta_j) \tag{1}$$

where E_i is the total emissions of a certain category; E_j is the emissions of the source j belonging to this category; T_j and δ_j represent the amount and coefficient of the source j, respectively. For the readers interested in the details of the emission coefficient, we list all the coefficients in Appendix A.

To facilitate the analysis, the estimated GHG emissions were boiled down into carbon equivalents, in line with Intergovernmental Panel on Climate Change (IPCC) [45], such that the GHG effect caused by 1 t N_2O and 1 t CH_4 is equivalent to that caused by 298 t CO_2 (81.2727 t C) and 25 t CO_2 (6.8182 t C), respectively [46].

2.1.2. Agricultural GHG Absorption

In terms of the agricultural GHG absorption, we considered the carbon dioxide that crops absorb by photosynthesis in a life cycle, whose calculation equation is:

$$C = \sum_{i=1}^{k} C_i = \sum_{i=1}^{k} c_i Y_i (1 - r_i)/HI_i, \tag{2}$$

where, C is the total amount of carbon dioxide absorbed by crops, C_i is the amount of carbon absorbed by a certain crop, and c_i is the corresponding coefficient of absorption, Y_i, r_i, HI_i is the economic yield, moisture content factor and economic coefficient of crop i, respectively. The corresponding coefficients can be referred to in Table A5 of Appendix A.

2.2. Spatial Correlation

To reveal the spatial correlation of agricultural net GHG emissions, Moran's I autocorrelation analysis was adopted in this paper. We provide the calculation of Moran's I in Appendix B.

Before calculating Moran's I, we needed to quantify the spatial relationship of geographic units by spatial weight matrix. When setting the matrix, related research often applies the "0–1" adjacency matrix, which only considers the adjacent relation but ignores the difference caused by geographical distance [47]. Instead, based on the distance attenuation function, we took the reciprocal of the distance square among geographic centers of 30 provinces as the spatial weight, so as to consider the possible interaction among provinces which were geographically close but not adjacent.

2.3. Convergence Test

2.3.1. α Convergence Test

α convergence reflects the deviation of regional agricultural net GHG emissions from the overall average level. In this paper, we explored the convergence or divergence of agricultural net GHG emissions by testing the α coefficient, which denoted the average deviation of each value from the mean. The equation is as:

$$\alpha = \sqrt{\left[\sum_i (\ln E_{it} - \overline{\ln E_t})^2\right]\Big/ N}$$ (3)

where, $\ln E_{it}$ is the logarithmic value of the agricultural net GHG emissions of the province i in period t, $\overline{\ln E_t}$ is the logarithmic value of the average emissions of all provinces during period t, N is the number of provinces. It was possible to calculate the α value for every year, and if the value gradually decreased with the passage of time, it suggested the agricultural net GHG emissions showed α convergence, otherwise, α convergence did not exist.

2.3.2. Conditional β Convergence Test

β convergence consists of absolute β convergence and conditional β convergence. Absolute β convergence is used to judge the relationship between the growth rate and the initial level of the net emissions, and if they were related negatively, it indicated that there was an absolute β convergence and a catch-up trend existed in backward areas. Different from absolute β convergence, conditional β convergence does not only take the primary level into account, but considers several factors that may also affect the convergence. To explore the influence of different factors, we selected the conditional β convergence test, which can be calculated by Equation (4):

$$\ln\left(\frac{E_{it}}{E_{i,t-1}}\right) = \alpha + \beta \ln(E_{i,t-1}) + \gamma X + \mu_i + \lambda_t + \varepsilon_{it}$$ (4)

where, E_{it} and $E_{i,t-1}$ are net agricultural GHG emissions in the t and $t-1$ years of the province i, respectively. X represents a set of control variables, and γ is the corresponding parameter; μ_i is the individual-fixed effect and λ_t is the time-fixed effect; ε_{it} is the error term. If β is negative and statistically significant, it indicates that negative correlation exists between the growth rate and the primary level, which means there is conditional β convergence, if not, conditional β convergence does not exist.

Due to the technology diffusion and the policy imitation of neighboring regions, there may have been spatial correlation of agricultural net GHG emissions. Therefore, when conducting conditional β convergence tests, it was necessary to adopt spatial econometric models, and we mainly considered the spatial lag model (SAR) and spatial error model (SEM), which can be estimated by Equations (5) and (6) respectively.

$$\ln\left(\frac{E_{it}}{E_{i,t-1}}\right) = \alpha + \rho W \ln\left(\frac{E_{it}}{E_{i,t-1}}\right) + \beta \ln(E_{i,t-1}) + \gamma X + c_i + \eta_t + \varepsilon_{it}$$ (5)

where, ρ is the spatial autoregressive coefficient, which measures the mutual influence of the growth rates of the net emissions among provinces. The meanings of other variables are the same as in Equation (4).

$$\ln\left(\frac{E_{it}}{E_{i,t-1}}\right) = \alpha + \beta \ln(E_{i,t-1}) + \gamma X + c_i + \eta_t + \mu_{it}, \ \mu_{it} = \lambda W \mu_{it} + \varepsilon_{it}$$ (6)

where, λ is the coefficient of spatial error, which is used to measure the spatial correlation degree of error terms. The meanings of other variables are consistent with Equation (4). Elhorst suggested the selection of models could refer to Lagrange multiplier (LM) test: if the LM test statistics (or robust LM test statistics) of spatial lag effect is more significant than that of spatial error effect, SAR should be selected; on the contrary, SEM is more suitable [48]. In addition, the model can be selected according to R^2, log likelihood, Akaike Information Criterion (AIC), Schwarz Criterion (SC) and other indicators.

In terms of variables, the logarithm of the growth rate of the net emissions was taken as the explained variable, and the logarithm of the net emissions was the core variable. Referring to the relevant studies [33–35], we chose six factors that may have affected the convergence of emissions as control variables, as shown in Table 1.

Table 1. Description of control variables for the conditional β convergence test.

Control Variable	Symbol	Calculation	Unit
Agricultural labor	*al*	Agricultural practitioners	10^4 capita
Multiple cropping	*mc*	Ratio of sown area to cultivated area of crops	-
Agricultural disaster	*ad*	Ratio of agricultural disaster area to crop sowing area	-
Agricultural economy	*ae*	Ratio of gross agricultural output value to rural population	10^4 CNY per capita
Industrial structure	*is*	Ratio of non-crop production value to gross agricultural output value	-
Fiscal expenditure	*fe*	Ratio of agricultural fiscal expenditure to total fiscal expenditure	-

Note: To guarantee the gross agricultural output value and the non-crop production value was comparable, it was necessary to convert the corresponding value of each year into the actual value calculated at the comparable price in 2007. To make sure the data was smooth, two variables, agricultural labor and agricultural economy, needed to be logarithmized.

2.4. Data Source

Excluding Hong Kong, Macao, Taiwan and Tibet due to missing data, the primary data covers 30 provinces of China from 2007 to 2016, consisting of all kinds of activity data involved in the calculation of agricultural GHG emissions and absorption, and the control variables of β convergence test. The data is introduced in Table 2.

Table 2. Introduction of data involved in the study.

Calculation	Category	Indicator	Source
Agricultural GHG emissions	Farmland utilization	Application amount of fertilizers, pesticides, plastic mulch, plowing area and irrigation area	*China Rural Statistical Yearbook*
	Rice planting	Planting area of rice in all provinces	*China Rural Statistical Yearbook and National Compilation of Cost and Income Information on Agricultural Products*
	Ruminant breeding	Year-end stock of cattle, horses, donkeys, mules, pigs, goats and sheep	*China Animal Industry Yearbook and Provincial Statistical Offices*
	Agricultural energy consumption	Amount of coal, coke, crude oil, gasoline, kerosene, diesel oil, fuel oil and natural gas used in agricultural production	*China Energy Statistics Yearbook and Provincial Statistical Offices*

Table 2. *Cont.*

Calculation	Category	Indicator	Source
Agricultural GHG absorption	Main crops	Yield of rice, wheat, corn, soybean, rapeseed, peanut, sunflower, cotton, potato, sugar cane, beet, vegetable, melon, tobacco and other crops	*China Rural Statistical Yearbook and National Compilation of Cost and Income Information on Agricultural Products*
Conditional β convergence test	Control variables	Crop sown area, cultivated land area, agricultural disaster area, gross agricultural output value, total rural population, non-crop production value, fiscal expenditure on agriculture and total fiscal expenditure	*China Rural Statistical Yearbook*

3. Empirical Analysis

3.1. Calculation and Analysis of Agricultural Net GHG Emissions

3.1.1. The Structure of Agricultural Net GHG Emissions

According to the method above, we estimated the net GHG emissions of agriculture in China from 2007 to 2016 and analyzed the structure, as Figure 1 exhibits.

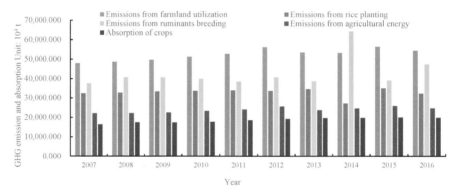

Figure 1. The histogram of agricultural net greenhouse gas (GHG) emissions' structure of China from 2007 to 2016.

Figure 1 presents the development of agricultural GHG emission and absorption in China. As for emission, the GHG emissions from farmland utilization averaged $52,383.423 \times 10^4$ t, accounting for 34.494% of the total GHG emission, which was the largest emissions source. While the source with the smallest emissions was the agricultural energy, with an average of $23,897.309 \times 10^4$ t, only accounted for 15.736% of the total. Besides, the mean of the other two sources were $32,850.876 \times 10^4$ t and $42,732.930 \times 10^4$ t separately, contributed 21.632% and 28.139% of the total amount. Observing the evolution of the structure, the variation trend of GHG emission sources was significantly different. Except the emissions from rice planting, which slightly decreased with an annual decline rate of −0.025%, those from farmland utilization, ruminant breeding and agricultural energy all performed a fluctuating rise, whose annual growth rate were 1.287%, 1.437%, 2.614%, respectively. As for crop GHG absorption, its average amount was $18,616.541 \times 10^4$ t, which rose stably during the research period with an annual growth rate of 2.129%. In the agricultural GHG emission structure of China, the amount of absorption was much lower than emissions.

3.1.2. Temporal Evolution of Agricultural Net GHG Emissions

To understand the temporal evolution of the net emissions, the corresponding box-plot was drawn (Figure 2).

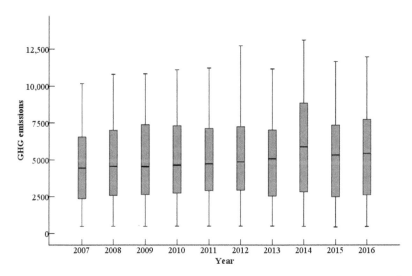

Figure 2. The box-plot of agricultural net GHG emissions of China from 2007 to 2016 (Unit: 10^4 t).

The average of provincial net emissions was 4999.916×10^4 t during 10 years. In 2007, the net emission was 4520.424×10^4 t, at the valley in 10 years, and then increased steadily, reaching the 10-year peak of 5842.196×10^4 t in 2014, declined in 2015 and recovered slightly in 2016, showing a fluctuating trend on the whole. Observing the extremum of agricultural net GHG emissions, there was a significant difference among provinces in different years. The minimum did not change significantly and stabilized at about 500.000×10^4 t, while the maximum experienced two light fluctuations, reaching the valley of $10,170.893 \times 10^4$ t in 2007 and the peak of $13,104.337 \times 10^4$ t in 2014. As can be seen from the interquartile range, the gap of the net emissions among provinces in China was gradually widening.

3.1.3. Spatial Distribution of Agricultural Net GHG Emissions

To show the spatial pattern of agricultural net GHG emissions, we divided the provincial average net emissions into five grades from high to low by natural breakpoint method, and the corresponding spatial distribution map was drawn (Figure 3).

Figure 3 shows the spatial distribution of the net emissions in China. From 2007 to 2016, there were four provinces with the highest emissions, namely Henan, Hunan, Shandong and Heilongjiang, where Henan was the only one whose emissions exceeded $10,000.000 \times 10^4$ t. There were seven provinces in the second grade, which were Sichuan, Jiangxi, Hubei, etc. The provinces with medium emission consisted of six provinces, including Guangdong, Yunnan, Jilin, etc. Then the next level included seven provinces, followed by Gansu, Guizhou, Zhejiang and so on. Last, six provinces belonged to lowest level, namely Qinghai, Hainan, Ningxia, etc. The regional pattern of the net emissions presented the "center–periphery" pattern described by New Economic Geography. Meanwhile, the provinces with high emissions were mainly agglomerated in the middle reaches of the Yangtze River, which had always been China's major agricultural production provinces and were also the central area of the agricultural GHG emissions. In addition, the provinces with low emissions were mainly located in the northwest region, whose agriculture were relatively underdeveloped and the GHG emissions caused by agriculture were naturally less. Apart from the two regions, other regions in China were characterized by staggered distribution of provinces at different GHG emission levels. As can be seen from Figure 3, the emissions showed certain agglomeration in space.

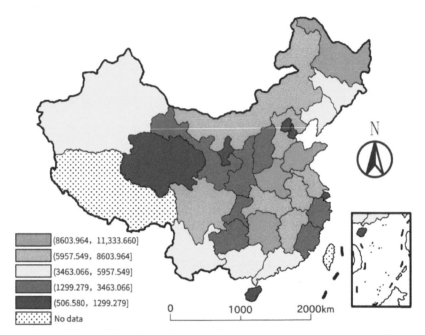

Figure 3. The spatial distribution of agricultural net GHG emissions in China (Unit: 10^4 t).

3.2. Spatial Correlation Analysis of Agricultural Net GHG Emissions

3.2.1. Global Correlation Analysis

By using Stata 15.0 software, the global Moran's I of the net emissions from 2007 to 2016 was calculated, and the results are shown in Table 3.

Table 3. The Moran's I of agricultural net GHG emissions from 2007 to 2016 in China.

Year	Moran's I	Z-Statistics	p-Value
2007	0.111	1.507	0.066
2008	0.118	1.583	0.057
2009	0.128	1.682	0.046
2010	0.122	1.625	0.052
2011	0.128	1.683	0.046
2012	0.171	2.141	0.016
2013	0.127	1.669	0.048
2014	0.147	1.875	0.030
2015	0.123	1.633	0.051
2016	0.133	1.732	0.042

The Moran's I were all over 0.100 and passed the Z-test at a significance level of 10%, which indicated that the agricultural net GHG emissions among provinces were not independent. Due to the similar natural resource, climate conditions and cropping structure, the net emissions became relevant in close provinces. In addition, the global Moran's I performed unstably, whose value gradually rose in the early years but decreased recently, showing an inverted U-shaped curve overall. Owing to the extensive mode of agricultural production and the single way of emission mitigation, the development of agricultural net GHG emissions behaved similarly among close provinces in the early years. However, with the promotion of environmental protection, various provinces had formulated diversified measures for emission reduction according to their own orientation, resulting in a gradual decrease of net emissions' correlation.

3.2.2. Local Correlation Analysis

It was hard to describe the specific spatial correlation by global Moran's I, so we took the 10-year average emissions of 30 provinces as the study objects, and chose the Moran scatter plot to investigate the local spatial correlation, as shown in Figure 4.

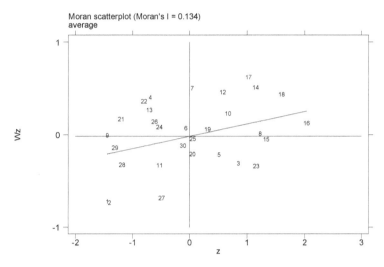

Figure 4. The local Moran scatter diagram of average agricultural net GHG emissions of 30 provinces. Note: 1 is Beijing, 2 is Tianjin, 3 is Hebei, 4 is Shanxi, 5 is Inner Mongolia, 6 is Liaoning, 7 is Jilin, 8 is Heilongjiang, 9 is Shanghai, 10 is Jiangsu, 11 is Zhejiang, 12 is Anhui, 13 is Fujian, 14 is Jiangxi, 15 is Shandong, 16 is Henan, 17 is Hubei, 18 is Hunan, 19 is Guangdong, 20 is Guangxi, 21 is Hainan, 22 is Chongqing, 23 is Sichuan, 24 is Guizhou, 25 is Yunnan, 26 is Shaanxi, 27 is Gansu, 28 is Qinghai, 29 is Ningxia and 30 is Xinjiang.

The Moran's I scatter plot included four quadrants, corresponding to "high–high" agglomeration (H–H), "low–high" agglomeration (L–H), "low–low" agglomeration (L–L) and "high–low" agglomeration (H–L) in turn. From Figure 4, the scatters in the first and third quadrants were more than those in the other two quadrants, i.e., the provinces belonging to the "high–high" and the "low–low" agglomeration took the majority. Among them, the provinces whose scatters were located the first quadrant totaled up to eight provinces, including Jilin, Jiangsu, Anhui, etc., showing the trend that high emission areas were surrounded by similar ones. The provinces belonging to the second quadrant included Shanxi, Liaoning, Fujian, etc., owning the character that low emission areas were surrounded by high emission areas. The provinces in the third quadrant comprised of Beijing, Tianjin, Shanghai and so on, performing that the low emission areas were surrounded by the similar ones. The provinces of the fourth quadrant included Hebei, Inner Mongolia, Sichuan and so on, which were characterized by high emission areas surrounded by those who performed oppositely. To conclude, the provinces of "high–high" agglomeration were mainly situated in the middle reaches of the Yangtze River, while those of "low–low" agglomeration mainly located in the northern coastal, eastern coastal and northwest regions. In addition, provinces around the country showed obvious polarization of high-value and low-value agglomeration.

Aiming to distinguish the agglomeration under the significance level of 5%, we selected the base year (2007), the last year (2016), the year with the strongest (2012) and the weakest spatial correlation (2007), and then drew the corresponding Lisa agglomeration maps, as shown in Figure 5.

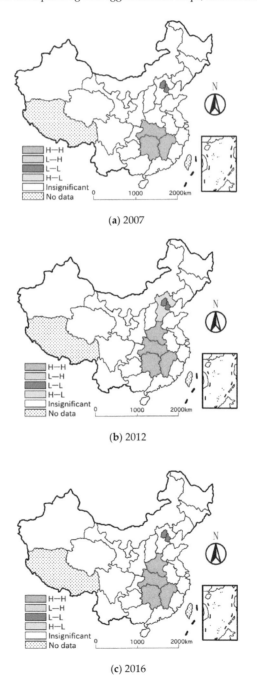

(a) 2007

(b) 2012

(c) 2016

Figure 5. Lisa agglomeration map of agricultural net GHG emissions in China in 2007, 2012 and 2016.

As seen from Figure 5, under the 5% significance level, the provinces showing significant agglomeration had decreased sharply. There were mainly two types of agglomeration left, as the net emissions in middle reaches of the Yangtze River mainly exhibited "high–high" agglomeration, and those in the northern coastal region appeared to show "low–low" agglomeration, indicating that an obvious polarization existed. With the passage of time, the spatial correlation pattern had changed greatly. Specifically, in 2007, the spatial correlation was the weakest. The "high–high" agglomeration areas included Jiangxi, Hubei and Hunan provinces, while Beijing and Tianjin showed "low–low" agglomeration. By 2012, Beijing and Tianjin continued to be the "low–low" agglomeration centers, and Henan became a new center of "high–high" agglomeration. Besides, Hebei, as a transitional area of two agglomerations, was significantly characterized by "high–low" agglomeration. By 2016, the pattern of "low–low" and "high–high" agglomeration had not changed, but the "high–low" agglomeration in Hebei was no longer significant.

3.3. Convergence Test

The analysis above suggested that there were apparent gaps of the net emissions in different provinces, but they showed local spatial correlation in space. Then, with the passage of time, would the gaps narrow naturally? We applied α convergence test and conditional β convergence test to explore the gaps of the net emissions among provinces.

3.3.1. α Convergence Test

The α coefficient of China's agricultural net GHG emissions was calculated year by year, and we drew a corresponding line chart, as shown in Figure 6.

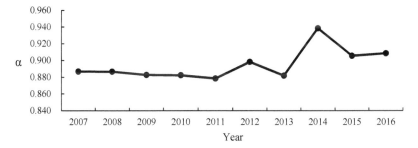

Figure 6. The α coefficient of agricultural GHG emissions of 30 provinces in China from 2007 to 2016.

Observing Figure 6, the α coefficient of the net emissions showed an overall upward tendency during the research period, indicating that there was no α convergence among 30 provinces. As the emissions developed diversely in different provinces, the gaps among provinces would continue to widen. Besides, the gaps diverged in different periods: the α coefficient decreased slowly from 0.887 to 0.879 in the first five years, showing weak convergence, while it presented fluctuating growth during the recent period, with a peak of 0.938 in 2014, which suggested that the evolution of agricultural net GHG emissions' gaps was unstable.

3.3.2. Conditional β Convergence Test

There was no α convergence for the national net emissions, which implied that the emission level of all provinces would not converge to a same stable standard. Then, we eased the convergence conditions and tried a conditional β convergence test, aiming to examine whether the "chase effect" of the emissions existed among provinces. The Moran's I estimation confirmed that emissions showed a positive spatial autocorrelation in provinces, so the conditional β convergence analysis should be conducted by the spatial econometric methods, instead of the ordinary panel econometric model.

In this paper, Stata 15.0 software was used for spatial econometric analysis, and we conducted spatial relevant tests for selecting spatial econometric models, which are shown in Table 4.

Table 4. The results of spatial panel econometric model test.

Test	χ^2	*p*-Value
LM Error (Burridge)	1.730	0.188
LM Error (Robust)	1.539	0.215
LM Lag (Anselin)	19.764	0.000
LM Lag (Robust)	19.572	0.000

As the Table 4 presents, the statistics of spatial lag effect were both significant at the 1% statistic level, while those of spatial error effect were not, so SAR was more suitable for convergence analysis. The data was a short panel of data, so there was no need for a panel unit root test or co-integration test. To ensure the robustness of the empirical results, the estimation of SEM and the ordinary panel econometric model was taken as the comparison. The Hausman statistics of the three models were significant at the 1% statistic level, all accepting the hypothesis of fixed effects. The regression results were exhibited in Table 5, where the log-likelihood ratio of SAR was 379.696, higher than that of SEM, and the R^2 was 0.518, better than the other two models, which also supported the rationality of SAR. Most of the variable coefficients of the three models presented the same sign and little numerical difference, indicating that the estimation results were robust.

Table 5. The results of conditional β convergence tested by three econometric methods.

Variables	SAR	SEM	Ordinary Panel Model
ln (*emissions*)	0.521 ***	0.737 ***	0.494 ***
	(3.57)	(3.06)	(6.93)
mc	−0.023	−0.023	−0.020
	(−1.11)	(−1.07)	(−0.90)
ln (*al*)	−0.231 ***	−0.312 **	−0.194 **
	(−2.89)	(−2.38)	(−2.58)
ad	−0.027	−0.048	−0.041
	(−0.77)	(−1.02)	(−1.23)
ln (*ae*)	−0.114 ***	−0.171 **	−0.123 ***
	(−2.80)	(−2.27)	(−4.88)
is	−0.042	−0.142	0.230
	(−0.21)	(−0.91)	(0.57)
fe	0.544	0.785	0.532
	(1.09)	(0.99)	(1.40)
ρ/λ	0.547	0.638	-
	(5.86)	(6.21)	-
Hausman	102.960	47.050	157.520
R^2	0.518	0.373	0.392
Log-likelihood	379.696	371.070	-

Note: Z-statistics are presented in the brackets; *** and ** mean variables were significant at 1% and 5% statistical levels, respectively.

Observing the estimation results of the SAR, we found the coefficient of emission's logarithm was positive at a 1% significance level, which proved the national agricultural net GHG emissions did not show conditional β convergence, that is, there was no "chase effect" among provinces. Due to the differences of resource endowments, economic development and industrial structure, not only did backward regions have higher net emissions than advanced regions in the primary stage, but the gap continued to widen over time. In addition, the spatial autoregressive coefficient ρ was 0.547 and passed the Z-test at the 1% significance level, implying the growth rates of the net emissions inclined to show spatial spillover effect significantly in close provinces.

As for the control variables, the coefficients of agricultural labor and agricultural economy were both negative and significant at the statistical level of 1%, indicating that the expansion of labor and the improvement of economy were conducive to lowering the growth rate of the net emissions. As the agricultural practitioners decrease, machines of irrigating and plowing would make up for the reduced labor in the agricultural production, suggesting that the GHG emissions caused by agricultural energy, such as diesel oil and agricultural gasoline, would increase. From the reality in China, with the acceleration of urbanization and more agricultural labor force transferring to the industry, the growth rate of agricultural net GHG emissions would be significantly increased. The development of agricultural economy also had a negative effect on the growth rate of net emissions. With the development of agricultural economy, green and clean mode of intensified production had been gradually replacing the extensive operation mode with excessive energy consumption and large emissions, which contributed to the emission mitigation.

4. Discussion

4.1. Implication

Agricultural GHG emissions plays an important role in global warming. As can be seen from the analysis, the situation of agricultural GHG in China is not optimistic, which reveals the urgency to accelerate the pace of emission mitigation.

(1) Although agricultural net GHG emissions in China had experienced some ups and downs in the sample years, it showed an overall upward trend. Moreover, based on the evolution, we could predict the net emissions may continue to grow in the next few years, manifesting the necessity to take measures for emissions mitigation. From the structure, GHG emissions from farmland utilization, ruminant breeding and agricultural energy all showed growing trends, especially those from farmland utilization, accounting for 34.494% of the total amount, should be paid more attention.

(2) Due to the obvious provincial difference, when formulating regional agricultural GHG emission mitigation policies, it is essential to establish differentiated emission reduction targets based on local conditions. The key is to apply the low-carbon development mode of farming: adjust agricultural production structure and plant low-emission and high-sink crops that adapt to local resources, increase investment in technology to enhance the efficiency of agricultural machinery and the utilization rate of energy, encourage agricultural practitioners to learn conservation-oriented fertilization techniques, use pesticides rationally and recycle the waste plastic mulch.

(3) Based on the results of two convergence tests, there is no convergence nationwide, so it is hard for the net emissions to reduce naturally. On the contrary, taking effective reduction measures is a possible way to bridge the provincial gaps. Besides, the net emissions showed spatial correlation, which interact and influence each other among provinces, suggesting that there is a possibility of regional cooperation. It is necessary for all regions to strengthen cooperation and share low-carbon technology, so as to cut down the net emissions coordinately through provincial correlation.

(4) A series of studies showed that some technologies may act in carbon sequestration and negative emission of carbon may be achieved [49]. Technologies and techniques as biochar [49], agroforestry systems [50] and conservational agriculture measurements [51,52] may act as additional benefit methods for reduce carbon emission in agriculture. At present, low-carbon technology is relatively insufficient for agriculture in China, and most of them are staying in the experimental stage. In the future, China should attach importance to the development of the technologies and techniques and apply them to practice as soon as possible.

4.2. Comparison

Comparing with the existing estimation of China, we find that difference in categories and sources of agricultural GHG leads to the diversity of the results. Taking Chen's estimation for example, he deems that China's agriculture served as a net GHG sink from 1991 to 2011 [30], while we believe that it was a GHG source from 2007 to 2016. Observing the calculation procedure, Chen calculates the GHG absorption from 15 kinds of crops, whose categories and corresponding coefficients are the same as our study, but there is a difference when calculating the emissions: Chen's research is on the basis of six sources of GHG emission source, inclusive of fertilizer, pesticide, plastic mulch, sheep, cow and pig, less than our study that considers 21 sources of GHG emissions. Therefore, there is little difference in the estimation of absorption between the two studies, but the emissions we estimated are much more than Chen's, consequently causing a significant difference in the net amount.

As for convergence, previous research focus on GHG emissions and relative indicators, and some scholars have agreed that the agricultural GHG emissions of China does not achieve convergence nationwide, neither does the intensity or performance [41,43], while the national agricultural carbon productivity has absolute β convergence [42]. Instead, we took the net emissions as the target, and found that neither of α convergence or conditional β convergence existed in the whole country, which is a complement to the existing research.

4.3. Improving Direction

It should be noted that there are some limitations of this study. First, it lacks the consideration for GHG effect of soil. The emission and sequestration of soil is closely related to the farming system, where the soil carbon loss caused by no-tillage, less tillage or complete cultivation often has a world of difference [34]. Some scholars adopt the test data of a certain region as GHG emission coefficient of soil, and applied it to the whole country [4,53]. However, China has a vast territory, and its geographical conditions and production patterns are quite different. Therefore, the application of such a simple way is unscientific. Based on the above reasons, GHG emission and sequestration of soil have not been considered in our measurement system of the agricultural GHG net emission.

Second, the impact of different farming methods was not considered. The agricultural practices, such as cover crops and straw returning, may also influence the carbon sequestration in the soil, resulting in a completely different net effect of GHG. As the major mode in China is smallholder production, it was hard to consider the influence of cultivation modes in different regions. For this reason, this paper did not consider different agricultural practices, which may have led to deviation in the result.

To refine the measurement system and ensure the accuracy of results, we will focus on taking the GHG effect of soil into research in the future. In addition, assessing the influence of different agricultural practices also becomes the next direction.

5. Conclusions

Based on the Moran's I, convergence tests and spatial econometric models, the study analyzed the spatial correlation and convergence of agricultural net GHG emissions in China. From 2007 to 2016, the average of the net emissions of all provinces was 4999.916×10^4 t, showing a fluctuating growth trend as a whole, and the gaps among provinces had been gradually widening. Most of the provinces with large emissions belonged to the middle reaches of the Yangtze River, while the provinces with low emissions were mainly located in the northwest of China. As for the spatial correlation, the global Moran's I of agricultural net GHG emissions was all over 0.100, implying the net emissions were spatially correlated, whose correlation level showed an inverted U-shaped curve as a whole. There was an obvious polarization of the net emissions, mainly exhibiting "high–high" and "low–low" agglomeration, whose agglomerating centers were situated in the middle reaches of the Yangtze River and the northern coastal region respectively. With the passage of time, the spatial correlation pattern had changed greatly. In terms of convergence, agricultural net GHG emissions did not show α convergence or conditional β convergence in the whole country. With time going by, the gaps among different provinces broadened, and there was no "chase effect" in the emissions' growth rate among provinces. In addition, the growth rate had a significant positive spatial spillover effect in close provinces, and the agricultural force and economic development had negative impact on the growth rate of the net emissions.

Author Contributions: Conceptualization, H.W. and W.C.; data curation, H.H. and H.W.; software, J.T. and H.H.; formal analysis, H.W.; writing and revision, all the authors; project administration, W.C.; funding acquisition, W.C.

Funding: This research was funded by a National Natural Science Foundation Youth Project (71704127), Major Social Science Planning Projects in Sichuan Province (SC17ZD06) and Sichuan Social Science Research "the 13th Five-Year Plan" (SC18TJ018).

Conflicts of Interest: The authors declare no conflict of interest.

Appendix A

We present the coefficients of different source of agricultural GHG emission and absorption in this section. The coefficients of farmland utilization are shown in Table A1.

Table A1. GHG emission coefficients of major sources in farmland utilization.

Farmland Utilization	Emission Coefficient
Fertilizer	$0.8956\ kg(C){\cdot}kg^{-1}$
Pesticide	$4.9341\ kg(C){\cdot}kg^{-1}$
Plastic mulch	$5.1800\ kg(C){\cdot}kg^{-1}$
Ploughing	$312.6000\ kg(C){\cdot}hm^{-2}$
Irrigation	$266.4800\ kg(C){\cdot}hm^{-2}$

Data source: Duan [29].

Due to the diverse climate and temperature, CH_4 emission rate in rice growth cycle was also various in different provinces. Table A2 presents the coefficients of rice cultivation in provinces.

Table A2. GHG emission coefficient of rice cultivation in provinces of China. Unit: $g(C){\cdot}m^{-2}$.

Province	Emission Coefficient	Province	Emission Coefficient
Beijing	13.23	Henan	17.85
Tianjin	11.34	Hubei	38.20
Hebei	15.33	Hunan	35.00
Shanxi	6.62	Guangdong	41.20
Neimenggu	8.93	Guangxi	36.40
Liaoning	9.24	Hainan	38.40
Jilin	5.57	Sichuan	16.90
Heilongjiang	8.31	Chongqing	16.90
Shanghai	31.26	Guizhou	16.10
Jiangsu	32.40	Yunnan	5.70
Zhejiang	35.60	Shanxi	12.51
Anhui	31.90	Gansu	6.83
Fujian	34.60	Qinghai	0.00
Jiangxi	42.20	Ningxia	7.35
Shandong	21.00	Xinjiang	10.50

Data source: Min [54].

The coefficients of major ruminants are shown in Table A3.

Table A3. GHG emission coefficient of major ruminants. Unit: kg(C)·head^{-1}·a^{-1}.

Ruminant	Emission Coefficient of Intestinal Fermentation	Emission Coefficient of Feces Management
Cow	395.5600	24.5520
Horse	122.7600	11.1848
Donkey	68.2000	6.1380
Mule	68.2000	6.1380
Pig	6.8200	27.2800
Goat	34.1000	1.1594
Sheep	34.1000	1.0230

Data source: IPCC [45].

Table A4 presents the coefficients of emissions from agricultural energy, which has been converted in line with China through mass conversion of the same calorific value.

Table A4. GHG emission coefficient of main agricultural energy consumption.

Agricultural Energy	Emission Coefficient	Agricultural Energy	Emission Coefficient
Coal	1.4676 kg(C)·kg^{-1}	Kerosene	4.5844 kg(C) kg^{-1}
Coke	2.9573 kg(C) kg^{-1}	Diesel	4.6031 kg(C)·kg^{-1}
Crude	4.3808 kg(C)·kg^{-1}	Fuel oil	4.6218 kg(C)·kg^{-1}
Gasoline	4.6939 kg(C)·kg^{-1}	Natural gas	2.9047×10^4 t(C)·m^{-3}

Data source: Guan [46].

In Table A5, we list the corresponding coefficient of main crops.

Table A5. Moisture content factor, carbon absorption rate and economic coefficient of main crops.

Crop	Moisture Content Factor	Carbon Absorption Rate	Economic Coefficient
Rice	0.12	0.414	0.45
Wheat	0.12	0.485	0.40
Core	0.13	0.471	0.40
Beans	0.13	0.450	0.34
Rapeseed	0.10	0.450	0.25
Peanut	0.10	0.450	0.43
Sunflower	0.10	0.450	0.30
Cotton	0.08	0.450	0.10
Yams	0.70	0.423	0.70
Sugarcane	0.50	0.450	0.50
Beet	0.75	0.407	0.70
Vegetables	0.90	0.450	0.60
Melons	0.90	0.450	0.70
Tobacco	0.85	0.450	0.56
Other crops	0.12	0.450	0.40

Data source: Wang [22].

Appendix B

In this section, we present the calculation of Moran's I for readers. The specific equation of global Moran's I is:

$$
I = \frac{n \sum\limits_{i=1}^{n} \sum\limits_{j=1}^{n} W_{ij}(x_i - \bar{x})(x_j - \bar{x})}{\sum\limits_{i=1}^{n} \sum\limits_{j=1}^{n} W_{ij} \sum\limits_{i=1}^{n} (x_i - \bar{x})} = \frac{\sum\limits_{i=1}^{n} \sum\limits_{j \neq 1}^{n} W_{ij}(x_i - \bar{x})(x_j - \bar{x})}{S^2 \sum\limits_{i=1}^{n} \sum\limits_{j \neq 1}^{n} W_{ij}}, \tag{A1}
$$

$$S^2 = \frac{1}{n}\sum_i (x_i - \overline{x})^2. \tag{A2}$$

In Equations (A1) and (A2), I is the global Moran's I; n is the number of spatial units; x is the net emissions of each space unit; W_{ij} is a spatial weight matrix. Moran's I ranges from -1 to 1, when Moran's I is a positive number at a given significance level, it proves that a positive correlation exist among the observed values. When Moran's I is negative, it indicates that there is a negative correlation. The closer the value is to 0, the weaker the correlation is. In addition, the Z-test is suitable for Moran's I statistical test.

On the basis of global spatial autocorrelation analysis, local Moran's I was able to reveal the spatial autocorrelation of neighboring provinces, whose calculation equation is:

$$I_i = \frac{n(x_i - \overline{x})\sum_j W_{ij}(x_j - \overline{x})}{\sum_i (x_i - \overline{x})^2} = \frac{nz_i \sum_j W_{ij}z_j}{z^T z} = z'_i \sum_j W_{ij}z'_j. \tag{A3}$$

In Equation (A3), I_i is the value of Moran's I; Z_i', Z_j' are observed values that are normalized by standard deviation; other variables have the consistent meaning to their counterpart in Equation (A1).

References

1. Johnson, M.F.; Franzluebbers, A.J.; Weyers, S.L.; Reicosky, D.C. Agricultural opportunities to mitigate greenhouse gas emissions. *Environ. Pollut.* **2007**, *150*, 107–124. [CrossRef] [PubMed]
2. Li, B.; Zhang, J.B.; Li, H.P. Research on spatial-temporal characteristics and affecting factors decomposition of agricultural carbon emission in China. *China Popul. Resour. Environ.* **2011**, *21*, 80–86. [CrossRef]
3. Tubiello, F.N.; Salvatore, M.; Golec, R.D.C.; Ferrara, A.; Rossi, S.; Biancalani, R.; Federici, S.; Jacobs, H.; Flammini, A. Agriculture, forestry and other land use emissions by sources and removals by sinks: 1990–2011 analysis. *FAO Stat. Div.* **2014**, *4*, 375–376.
4. Tian, Y.; Zhang, J.B. Study on the differentiation of net carbon effect in agricultural production in China. *J. Nat. Resour.* **2013**, *28*, 1298–1309. [CrossRef]
5. Tongwane, M.; Mdlambuzi, T.; Moeletsi, M.; Tsubo, M.; Mliswa, V.; Grootboom, L. Greenhouse gas emissions from different crop production and management practices in South Africa. *Environ. Dev.* **2016**, *19*, 25–35. [CrossRef]
6. Goglio, P.; Smith, W.N.; Grant, B.B.; Desjardins, R.L.; Gao, X.; Hanis, K.; Tenuta, M.; Campbell, C.A.; McConkey, B.G.; Nemecek, T.; et al. A comparison of methods to quantify greenhouse gas emissions of cropping systems in LCA. *J. Clean. Prod.* **2017**, *172*, 4010–4017. [CrossRef]
7. Zhang, X. Multiple cropping system expansion: Increasing agricultural greenhouse gas emissions in the north China plain and neighboring regions. *Sustainability* **2019**, *11*, 3941. [CrossRef]
8. Lu, X.H.; Kuang, B.; Li, J.; Han, J.; Zhang, Z. Dynamic evolution of regional discrepancies in carbon emissions from agricultural land utilization: Evidence from Chinese provincial data. *Sustainability* **2018**, *10*, 552. [CrossRef]
9. Lin, H.W.; Jin, Y.F.; Giglio, L.; Foley, J.A.; Randerson, J.T. Evaluating greenhouse gas emissions inventories for agricultural burning using satellite observations of active fires. *Ecol. Appl.* **2012**, *22*, 1345–1364. [CrossRef]
10. Wang, W.W.; Zhang, J.B.; Wang, P.C. Carbon emission measurement using different utilization methods of waste products: Taking cotton straw resources of south Xinjiang in China as an example. *Nat. Environ. Pollut. Technol.* **2018**, *17*, 383–390.
11. Kipling, R.P.; Taft, H.E.; Chadwick, D.R.; Styles, D.; Moorby, J. Challenges to implementing greenhouse gas mitigation measures in livestock agriculture: A conceptual framework for policymakers. *Environ. Sci. Policy* **2019**, *92*, 107–115. [CrossRef]
12. Garnier, J.; Noë, L.J.; Marescaux, A.; Cobena, A.S.; Lassaletta, L.; Silvestre, M.; Thieu, V.; Billen, G. Long-term changes in greenhouse gas emissions from French agriculture and livestock (1852–2014): From traditional agriculture to conventional intensive systems. *Sci. Total Environ.* **2019**, *660*, 1486–1501. [CrossRef] [PubMed]

13. Bellarby, J.; Tirado, R.; Leip, A.; Weiss, F.; Lesschen, J.P.; Smith, P. Livestock greenhouse gas emissions and mitigation potential in Europe. *Glob. Chang. Biol.* **2013**, *19*, 3–18. [CrossRef] [PubMed]
14. Smith, W.N.; Grant, B.B.; Desjardins, R.L.; Worth, D.; Huffman, E.C. A tool to link agricultural activity data with the DNDC model to estimate GHG emission factors in Canada. *Agric. Ecosyst. Environ.* **2010**, *136*, 301–309. [CrossRef]
15. Thamo, T.; Kingwell, R.S.; Pannell, D.J. Measurement of greenhouse gas emissions from agriculture: Economic implications for policy and agricultural producers. *Aust. J. Agr. Resour. Ec.* **2013**, *57*, 234–252. [CrossRef]
16. Huang, K.-T.; Wang, J.C. Greenhouse gas emissions of tourism-based leisure farms in Taiwan. *Sustainability* **2015**, *7*, 11032–11049. [CrossRef]
17. Liski, J.; Perruchoud, D.; Karjalainen, T. Increasing carbon stocks in the forest soils of Western Europe. *Ecol. Manag.* **2002**, *169*, 159–175. [CrossRef]
18. Baritz, R.; Seufert, G.; Montanarella, L.; Ranst, E.V. Carbon concentrations and stocks in forest soils of Europe. *Ecol. Manag.* **2010**, *260*, 262–277. [CrossRef]
19. Pan, Y.; Birdsey, R.A.; Fang, J.; Houghton, R.; Kauppi, P.E.; Kurz, W.A.; Philips, O.L.; Shvidenko, A.; Lewis, S.L.; Canadell, J.G.; et al. A large and persistent carbon sink in the world's forests. *Science* **2011**, *333*, 988–993. [CrossRef]
20. Erb, K.H.; Kastner, T.; Luyssaert, S.; Houghton, R.A.; Kuemmerle, T.; Olofsson, P.; Haberl, H. Bias in the attribution of forest carbon sinks. *Nat. Clim. Chang.* **2013**, *3*, 854–856. [CrossRef]
21. Luyssaert, S.; Schulze, E.D.; Börner, A.; Knohl, A.; Hessenmöller, D.; Law, B.E.; Ciais, P.; Grace, J. Old-growth forests as global carbon sinks. *Nature* **2008**, *455*, 213–215. [CrossRef] [PubMed]
22. Wang, X.L. *Carbon Dioxide, Climate Change, and Agriculture*; China Meteorological Press: Beijing, China, 1996.
23. Han, Z.Y.; Meng, Y.L.; Xu, J.; Wu, Y.; Zhou, Z.G. Temporal and spatial difference in carbon footprint of regional farmland ecosystem-taking Jiangsu Province as a case. *J. Agro Environ. Sci.* **2012**, *5*, 1034–1041.
24. West, T.O.; Marland, G. A Synthesis of carbon sequestration, carbon emissions, and net carbon flux in agriculture: Comparing tillage practices in the United States. *Agric. Ecosyst. Environ.* **2002**, *91*, 217–232. [CrossRef]
25. Vleeshouwers, L.M.; Verhagen, A. Carbon emission and sequestration by agricultural land use: A model study for Europe. *Glob. Change Biol.* **2010**, *8*, 519–530. [CrossRef]
26. Wear, D.N.; Coulston, J.W. From sink to source: Regional variation in U.S. forest carbon futures. *Sci. Rep.* **2015**, *5*, 16518. [CrossRef] [PubMed]
27. Popp, M.; Nalley, L.; Fortin, C.; Smith, A.; Brye, K. Estimating net carbon emissions and agricultural response to potential carbon offset policies. *Agron. J.* **2011**, *4*, 1131–1143. [CrossRef]
28. Xiong, C.H.; Yang, D.G.; Huo, J.W.; Wang, G.L. Agricultural net GHG Effect and agricultural carbon sink compensation mechanism in Hotan prefecture, China. *Pol. J. Environ. Stud.* **2017**, *1*, 365–373. [CrossRef]
29. Duan, H.P.; Zhang, Y.; Zhao, J.B.; Bian, X.M. Carbon footprint analysis of farmland ecosystem in China. *J. Soil Water Conserv.* **2011**, *25*, 203–208. [CrossRef]
30. Chen, L.Y.; Xue, L.; Xue, Y. Spatial agglomeration and variation of China's agricultural net carbon sink. *J. Ecol. Environ.* **2015**, *11*, 1777–1784.
31. Neumayer, E. Can natural factors explain any cross-country differences in carbon dioxide emissions? *Energy Policy* **2002**, *30*, 7–12. [CrossRef]
32. Lantz, V.; Feng, Q. Assessing income, population, and technology impacts on CO_2 emissions in Canada: Where's the EKC? *Ecol. Econ.* **2006**, *57*, 229–238. [CrossRef]
33. Uchiyama, K. *Environmental Kuznets Curve Hypothesis and Carbon Dioxide Emissions*; Springer: Tokyo, Japan, 2016.
34. Lal, R. Carbon emission from farm operations. *Environ. Int.* **2004**, *30*, 981–990. [CrossRef] [PubMed]
35. Zhang, J.B.; Cheng, L.L.; He, K. The difference of China's agricultural low-carbon economic efficiency in spatial and temporal and its influencing factors: A perspective of carbon input. *J. Environ. Econ.* **2017**, *2*, 36–51. [CrossRef]
36. Cole, C.V.; Duxbury, J.; Freney, J.; Heinemeyer, O.; Minami, K.; Mosier, A.; Paustian, K.; Rosenberg, N.; Sampson, N.; Sauerbeck, D.; et al. Global estimates of potential mitigation of greenhouse gas emissions by agriculture. *Nutr. Cycl. Agroecosyst.* **1997**, *49*, 221–228. [CrossRef]

37. Dace, E.; Muizniece, I.; Blumberga, A.; Kaczala, F. Searching for solutions to mitigate greenhouse gas emisons by agricultural policy decisions—Application of system dynamics modeling for the case of Latvia. *Sci. Total Environ.* **2015**, *527*, 80–90. [CrossRef]

38. Strazicich, M.C.; List, J.A. Are CO_2 emission levels converging among industrial countries? *Environ. Resour. Econ.* **2003**, *24*, 263–271. [CrossRef]

39. Westerlund, J.; Basher, S.A. Testing for convergence in carbon dioxide emissions using a century of panel data. *Environ. Resour. Econ.* **2008**, *40*, 109–120. [CrossRef]

40. Lee, C.C.; Chang, C.P. New evidence on the convergence of per capita carbon dioxide emissions from panel seemingly unrelated regressions augmented Dickey-Fuller tests. *Energy* **2008**, *33*, 1468–1475. [CrossRef]

41. Yang, X.Y. Regional differences and convergence of agricultural carbon emissions in China. *Hubei Agric. Sci.* **2016**, *55*, 1066–1072. [CrossRef]

42. Cheng, L.L.; Zhang, J.B.; Zeng, Y.M.; Wu, X.R. Analysis on the dynamic evolution and spatial club convergence of national agricultural carbon productivity. *J. Chin. Agric. Univ.* **2016**, *21*, 121–132. [CrossRef]

43. Wu, H.Y.; He, Y.Q.; Chen, R. Agricultural carbon emissions performance and its stochastic convergence in China-Based on SBM-Undesirable model and panel unit root test. *Chin. J. Ecoagri.* **2017**, *25*, 1381–1391. [CrossRef]

44. Anselin, L. *Spatial Econometrics: Methods and Models*; Kluwer Academic Publishers: Dordrecht, The Netherlands, 1988.

45. IPCC. *Climate Change 2007: The Physical Science Basis: Working Group I Contribution to the Fourth Assessment Report of the Intergovernmental Panel on Climate Change*; Cambridge University Press: New York, NY, USA, 2007.

46. Guan, X.L.; Zhang, J.B.; Wu, X.R.; Cheng, L.L. The shadow prices of carbon emissions in China's planting industry. *Sustainability* **2018**, *10*, 753. [CrossRef]

47. Wu, X.R.; Zhang, J.B.; Cheng, L.L.; Tian, Y. Potential of agricultural carbon reduction under climate change and its spatial correlation characteristics in China: Based on the spatial Durbin model. *China Popul. Resour. Environ.* **2015**, *25*, 53–61.

48. Elhorst, J.P. *Spatial Econometrics: From Cross-Sectional Data to Spatial Panels*; Springer: Berlin, Germany, 2014.

49. Smith, P. Soil carbon questionar and biochar as negative emission technologies. *Glob. Chang. Biol.* **2016**, *22*, 1315–1324. [CrossRef]

50. Albrecht, A.; Kandji, S.T. Carbon sequestration in tropical agroforestry systems. *Agric. Ecosyst. Environ.* **2003**, *99*, 15–27. [CrossRef]

51. Powlson, D.S.; Stirling, C.M.; Thierfelder, C.; White, R.P.; Jat, M.L. Does conservation agriculture deliver climate change mitigation through soil carbon sequestration in tropical agro-ecosystems? *Agric. Ecosyst. Environ.* **2016**, *220*, 164–174. [CrossRef]

52. Corbeels, M.; Cardinael, R.; Naudin, K.; Guibert, H.; Torquebiau, E. The 4 per 1000 goal and soil carbon storage under agroforestry and conservation agriculture systems in sub-Saharan Africa. *Soil Till. Res.* **2019**, *188*, 16–26. [CrossRef]

53. Tan, Q. Greenhouse gas emission in China's agriculture: Situation and challenge. *China Popul. Resour. Environ.* **2011**, *21*, 69–75. [CrossRef]

54. Min, J.S.; Hu, H. Calculation of greenhouse gases emission from agricultural production in China. *China Popul. Resour. Environ.* **2012**, *22*, 21–27.

Article

Economic Impacts of a Low Carbon Economy on Global Agriculture: The Bumpy Road to Paris [†]

Hans Jensen [1,*], Ignacio Pérez Domínguez [1,*], Thomas Fellmann [1], Paul Lirette [2], Jordan Hristov [1] and George Philippidis [1,3]

[1] Joint Research Centre, European Commission, 41092 Seville, Spain; thomas.fellmann@ec.europa.eu (T.F.); jordan.hristov@ec.europa.eu (J.H.); george.philippidis@ec.europa.eu (G.P.)
[2] Research and Analysis Directorate, Agriculture and Agri-Food Canada, Ottawa, ON K1A 0C5, Canada; paul.lirette2@canada.ca
[3] Aragonese Agency for Research and Development (ARAID), Centre for Agro-Food Research and Technology, Instituto Agroalimentario de Aragón (IA2), Government of Aragón, 50059 Zaragoza, Spain
[*] Correspondence: hans.jensen@ec.europa.eu (H.J.); ignacio.perez-dominguez@ec.europa.eu (I.P.D.)
[†] Disclaimer: The views expressed are purely those of the authors and may not in any circumstances be regarded as stating an official position of the European Commission and Agri-Food Canada. The results of the analysis based on the use of the Aglink-Cosimo model are outside the responsibility of the OECD and FAO Secretariats.

Received: 7 March 2019; Accepted: 16 April 2019; Published: 19 April 2019

Abstract: Limiting climate change below a 2 °C temperature increase this century will require substantial reductions of greenhouse gas emissions and the transition to a climate-friendly, low carbon society. In this paper, the economic impact of a less carbon-intensive economy on agricultural markets is estimated by means of an integrated modelling framework. First, the macroeconomic impacts of moving into a global low carbon economy are analysed by applying different carbon taxes in a general equilibrium modelling framework. Second, the potential adoption of emission mitigation technologies is quantified and used in the Aglink-Cosimo model to assess the impacts on agricultural markets of emission mitigation scenarios compatible with the 2.0 °C target prescribed in the Paris Agreement. Results for 2030 show reductions in global non-CO_2 GHG emissions from agriculture (i.e., methane and nitrous oxide) by 10, 16 and 19% in 50, 100 and 150 USD/t CO_2eq global carbon tax scenarios, respectively (Least Developed Countries excluded). Only between 0.6% and 1.3% of the global reduction is caused by indirect macroeconomic effects, although at the regional level they can cause up to 5.8% of the reduction in agricultural emissions. Results suggest that ambitious mitigation targets can provoke significant negative impacts on agricultural production and underline the importance of integrating GHG emission developments and impacts of related policies into agricultural market projections.

Keywords: Paris Agreement; climate change mitigation; agricultural sector; market outlook; Aglink-Cosimo

1. Introduction

The 21st Conference of the Parties of the United Nations Framework Convention on Climate Change (UNFCCC) in December 2015 resulted in the Paris Agreement, where parties agreed to take action to limit global temperature rise this century to well below 2 °C above pre-industrial levels [1], often referred as the "2 °C target". Limiting climate change below 2 °C will require substantial reductions of greenhouse gas (GHG) emissions and the transition to a climate-friendly, low carbon economy. The European Commission's report "Global Energy and Climate Outlook, Road from Paris" [2], provides an initial estimate of potential emission reductions by sectors in the global economy that are required to reach the 2 °C target. This estimation is done by comparing a business as usual reference scenario to a 2 °C target scenario for the world (Figure 1). The estimation indicates that the GHG emissions reduction required by 2030 could be achieved by the power sector (contributing 39% to the total mitigation effort), followed by "other energy" sectors (19%), industry (18%), agriculture (10%), buildings (6%), transport (4%) and waste (4%). These results exclude emissions and sinks for the "Land Use, Land Use Change and Forestry" sector (LULUCF). More precisely, global GHG emissions from the agriculture sector (i.e., only accounting for the agricultural non-CO_2 emissions methane and nitrous oxide) are estimated to rise to 6.283 gigatonnes of carbon dioxide equivalents (GtCO_2eq) by 2030 in the reference scenario, whereas they decline to 4.996 Gt CO_2eq in the 2 °C scenario. This represents a 20% reduction in global agricultural sector emissions by 2030 [3].

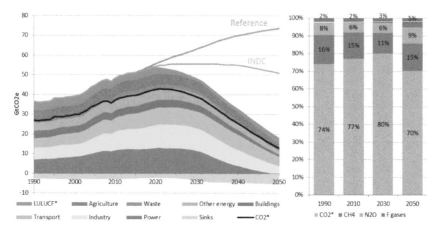

Figure 1. World GHG emissions in the 2030 reference, Intended Nationally Determined Contributions (INDCs) and 2 °C scenarios by sector (left) and by GHG (right). Source: [2]. Note: * CO_2 sinks are singled out and, therefore, not included in the LULUCF and CO_2 categories.

Other model simulations identify similar reduction targets for agricultural non-CO_2 emissions necessary to meet the objectives of the Paris Agreement. For example, the Integrated Model to Assess the Global Environment (IMAGE), the Global Change Assessment Model (GCAM) and the Model for Energy Supply Strategy Alternatives and their General Environmental Impact (MESSAGE) [4] calculated the need of global agricultural non-CO_2 emissions mitigation in the range of 11–18% by 2030 compared to the reference emissions (a reduction of 0.92–1.37 GtCO_2eq per year). The estimations in Kitous et al. [2] and Wollenberg et al. [4] are only two examples showing that the agricultural sector will be impacted both directly and indirectly by a low carbon economy. On the one hand, several studies point out that the agricultural sector has to directly contribute to emission reductions if the global climate change goals are to be met. This contribution has to come through direct emission reductions but also from increased land-use-based carbon dioxide removal [2,5–9], which will have a direct impact on agricultural production [10–13]. On the other hand, the agricultural sector will also be indirectly affected, as agricultural intermediate prices respond to the new economic environment. Given these foreseeable challenges, there is a need to adjust existing modelling tools, and eventually develop new ones, capable of analysing the economic impacts of a low carbon economy on agricultural markets in detail.

A variety of agricultural economic models are already equipped and utilized for the analysis of climate change mitigation on the agricultural sector [11–13]. However, the Aglink-Cosimo model [14,15], as one of the main partial equilibrium agro-economic models used to prepare medium-term agricultural market outlooks [16,17], is not yet prepared with all necessary features to account for agricultural emissions and respective mitigation efforts. Given that the agricultural projections produced annually by the OECD and FAO with the Aglink-Cosimo model establish the benchmark for many other agricultural economic models, it is specifically important that Aglink-Cosimo is able to transmit and measure the impact of a less carbon intensive economy on agricultural markets. Moreover, Aglink-Cosimo has important features that make it particularly suitable for analysing impacts on the agricultural sector of policies related to a movement towards a low carbon economy. For example, the model has a global coverage of the main agricultural commodities produced, consumed and traded, a detailed representation of domestic and trade-related agricultural policies, and accounts for substitution effects between agricultural commodities through explicit domestic price transmission equations [10,11]. Accordingly, enabling Aglink-Cosimo to transmit and measure the impact of a less carbon-intensive economy on agricultural markets is a major contribution to the future analysis of agricultural emission pathways and related impacts on agricultural market developments.

In this paper, the model adjustments necessary to enable Aglink-Cosimo to account for non-CO_2 emissions and to reflect the impacts of a low carbon economy are briefly outlined. This updated model is then used to simulate the economic impacts on agricultural markets of a global 2 °C target that is compatible with the Paris Agreement. Since Aglink-Cosimo is a partial equilibrium model, this scenario analysis requires first capturing the macro-economic impacts in a general equilibrium model framework and transmitting these changes to the Aglink-Cosimo model. In a second step, the agricultural sector's possible contribution to reductions in GHG emissions is analysed by implementing scenarios with a global GHG emission tax compatible with a 2 °C target. In addition, marginal abatement cost curves (MACC) for the main agricultural methane and nitrous oxide emission sources are introduced to capture the potential effects of technology development for mitigation (see methodological approach in Frank et al. [18]). This highlights the importance of technological progress for achieving a certain agro-environmental target, which is often neglected in the literature.

2. A Partial Equilibrium Modelling Framework

The Aglink-Cosimo model is a recursive-dynamic, partial equilibrium, multi-commodity market model of world agriculture. The model was developed by the OECD and FAO secretariats, with the double purpose of preparing medium-term (usually about 10 years) agricultural market outlooks [16,17], and to provide an economic simulation model for the assessment of policies [19–21] and economic changes related to the agricultural sector [22,23]. The model endogenously calculates the development of annual supply, demand and prices for the main agricultural commodities produced, consumed and traded worldwide. The present version of the model covers 82 individual countries and regions, 93 commodities and 40 world market clearing prices. Country and regional modules are developed and maintained by the OECD and FAO Secretariats, with important input in terms of data and analysis from country experts and national administrations. In a joint publication, the OECD and FAO provide annually a global outlook for the development of agricultural markets and prices. A large amount of expert knowledge is applied at various stages of the outlook process and Aglink-Cosimo is used to facilitate the consistent integration of this information from a markets intelligence perspective. Moreover, the outlook is built on the basis of specific assumptions on the short- and medium-term development of key macro-economic indicators (such as GDP, exchange rates, population, inflation and energy prices), which seem plausible at the moment of preparing the projections, given the current global environment [14,15]. For this paper, the model version released by the OECD-FAO with their 2017 agricultural market outlook was used. It includes market projections up to the year 2030 and a complete land allocation system introduced for 14 countries (Australia, Canada, Switzerland, Japan, South Korea, Mexico, Norway, New Zealand, United States of America, European Union, Argentina, Brazil, China and Russian Federation), taking into account double cropping systems in China, Brazil [24] and the United States [16]. Taking this initial model version and its underlying database, the following elements were added to analyse the impact of a less carbon intensive economy on the agricultural sector: (1) an enhanced land allocation system, (2) diminishing food demand elasticities with growing income, (3) increasing factor productivity and long run yield elasticities,

(4) a module on GHG emission accounting (i.e., estimation of emission intensities per agricultural production activity), and (5) incorporation of technological progress for emission abatement (i.e., decomposition of technological, production and structural emission reduction effects). These model improvements are briefly outlined below.

First, a complete land allocation system was imposed for all single developing countries (*Algeria, Angola, Bangladesh, Burkina Faso, Cambodia, Cameroon, Chad, Chile, Colombia, Congo, Côte d'Ivoire, Democratic Republic of the Congo, Egypt, Ethiopia, Gabon, Ghana, Haiti, India, Indonesia, Iran, Iraq, Israel, Kazakhstan, Kenya, Lao People's Democratic Republic, Lebanon, Libyan Arab Jamahiriya, Madagascar, Malawi, Malaysia, Mali, Mauritania, Morocco, Mozambique, Myanmar, Nigeria, Pakistan, Paraguay, Peru, Philippines, Rwanda, Saudi Arabia, Senegal, Somalia, South Africa, Sudan, Thailand, Tunisia, Turkey, Uganda, Ukraine, United Republic of Tanzania, Uruguay, Viet Nam, Yemen, Zambia and Zimbabwe*) and developing country regions (*Other Sub-Saharan Africa, Least Developed Countries (LDC), Subsaharan Africa, Other Asia Developed, Other Asia, LDC Asia, Other Oceania, LDC Oceania, Other South America and Caribbean, Other Middle East, Other Western Europe, and Other Eastern Europe*) specified in the model, where initially "pasture land" and "other crop land" (i.e. aggregate of the land used by all other crops not specifically included in Aglink-Cosimo) were exogenous. Incorporating a full land allocation system in the Aglink-Cosimo model is especially important in the context of emissions related to land use and land use change (LULUC). For this purpose, a full matrix of supply elasticities for crop land was estimated, specifically including pasture and other crop land, which, for example, allowed accounting for ruminant production returns on land allocation. Even though CO_2 emissions related to land use changes are not yet considered in the emission accounting of the model, capturing changes in total land use gives an indication of the effects that policy changes can have on GHG emission developments.

Second, adjustments to the income, own food and cross food demand elasticities were made in the model for developing countries. In particular, these elasticities were transformed to become variables (as opposed to constants), allowing them to decrease in value as wealth increases over time. This adjustment, thus, enables the ability for developing countries to close income gaps with developed economies (i.e., allows developing countries to move along the Engel curve).

Third, another important issue to consider in medium- to long-term analysis is factor productivity, which is expected to increase over time. Therefore, a long-term crop yield response to movements in agricultural commodity prices and input costs, as well as to the share of labour. was also introduced into the model. This adjustment included: (i) the estimation of long-term elasticities responding to historical long-term crop prices and cost signals [25], (ii) changes in the share of labour in the total cost index, following changes in real GDP per capita [16], and (iii) a new input demand system, reflecting that the move to a low carbon economy will likely affect the prices of fertilisers, chemicals and energy, which in turn could lead to changes in the input mix.

Fourth, modelling the contribution of the agricultural sector to GHG emission reduction targets involves calculating GHG emissions per agricultural production activity (i.e., emission intensities) and allowing the model to react when GHG emission mitigation policies, such as carbon taxes, are imposed. Therefore, the model was improved to account for the agricultural non-CO_2 emissions methane and nitrous oxide, which the UNFCCC attributes to the sector "agriculture", differently than CO_2 emissions and removals (LULUCF sector) and CO_2 emissions related to energy consumption at the farm and the processing of agricultural inputs (other sectors). These were calculated following the IPCC [26] guidelines at the tier 1 level and using FAOSTAT data for the emission factors [27–29]. GHG emissions were then calculated in the model per country or region by multiplying the activity data (i.e., hectares of land and heads of livestock) by the calculated emission factors. In order to perform different policy scenarios, the non-CO_2 emission inventories in Aglink-Cosimo were aggregated in CO_2 equivalents.

The calculation of emission factors was based on historical emissions and production data from FAOSTAT, but in order to allow for emission efficiency improvements reflecting the dynamics of production systems, trend functions were estimated. These trend functions for emission intensities were estimated within a robust Bayesian estimation framework that combined data from FAOSTAT on production quantities and emission inventories. The approach is further outlined in Jansson et al. [30], Pérez Domínguez et al. [31,32], and Van Doorslaer et al. [33]. Regarding carbon taxes, the taxes on emissions were introduced in the individual area harvested and livestock production equations, which allowed the analysis of tax effects in terms of emission reductions and production impacts at the individual country level. The carbon tax was introduced on a "per tonne of carbon-equivalent" basis and was applied to each production activity in each region captured by the model, so that emission intensity across activities and regions was taken into account.

Finally, technological (i.e., technical and management-based) mitigation options are incorporated into the analysis in the (reduced) form of regional marginal abatement cost curves (MACC) for different agricultural non-CO_2 emissions. These MACC are estimated ex-post based on information from Lucas et al. [34] and are depicted as the exponential function of the maximum potential degree of abatement given a certain carbon tax (i.e., the maximum emission reduction level to be reached when the cost of reducing the last tonne of emissions equals the price of the tax). With this it is possible to further disaggregate the changes in emissions and production related to different carbon taxes into: (a) production effects (i.e., reducing agricultural production), (b) structural effects (i.e., structural change in the agricultural sector due to trade or shifts in consumption preferences for agricultural commodities) and (c) technological effects (i.e., technological progress at the agricultural production level) [18].

3. Scenario Narratives and Design

This paper assesses the impact of a low carbon economy on the agricultural sector and focus on the potential contribution of the agricultural sector to global GHG emission reduction targets by means of global carbon tax scenarios. Currently, a large share of agricultural non-CO_2 GHG emissions stem from bovine meat and dairy production. In the past, GHG intensities from these livestock production activities have been reduced due to the evolution from less to more intensive productions systems, resulting in increases in commodity output per animal that are larger than the corresponding increases in emissions per animal [28]. Similarly, agricultural yields have evolved towards more intensive and resilient crop production systems. Taking these past trends into account, the option of retiring land from agricultural production, creating potential carbon sinks, is one possible strategy to reduce CO_2 emissions. This could be combined with changes in consumer's preferences towards diets containing less animal protein [35]. A way to accomplish this strategy and to enforce the contribution of the agricultural sector to GHG emission mitigation is to introduce a carbon tax per tonne of GHG emissions. This would effectively target commodities with higher GHG intensities, which typically would be ruminant meat and milk from less intensive livestock productions systems. The resulting commodity price increase would give an incentive to consumers to change their consumption habits to less emission intensive products (e.g., eating less beef).

Following this underlying narrative, three global carbon tax scenarios are tested against a business-as-usual medium-term reference situation without a carbon tax (baseline). In the carbon tax scenarios (Tax50, Tax100 and Tax150) the macroeconomic effects inherent in moving to a global low carbon economy are specifically accounted for. Moreover, the potential incorporation over time of new mitigation technologies linked to the carbon price scenarios are taken into account. In practice three separate homogenous carbon price paths for all countries, with the exception of Least Developed Countries (LDCs), are introduced, with carbon prices gradually increasing from 0 in 2020 to, respectively 50, 100 and 150 USD/t CO_2eq in 2030. With this scenario setting the impact that emission mitigation policies could have on agricultural production and consumer diets can also be highlighted. Mitigation policies in LDCs are not simulated, such as to avoid potential negative effects on regional production, aggravating food insecurity.

Given that Aglink-Cosimo is a partial equilibrium model, the total impact of a low carbon economy cannot be directly evaluated. The majority of emission reductions will have to be made by other sectors of the economy [2], and imposing a carbon tax on the global economy will induce macroeconomic effects (e.g., changes in prices for crude oil, fertilisers and pesticides, as well as changes in real GDP) that in turn will impact the agricultural sector. As macroeconomic variables are exogenous in the Aglink-Cosimo model, the macroeconomic impact of a low carbon economy has to be first captured and quantified in a Computable General Equilibrium (CGE) model and then transmitted to the agricultural economic model. For this, a set of carbon tax scenarios is simulated using the Modular Applied GeNeral Equilibrium Tool (MAGNET) model and the GTAP database version 9 with base year 2011 [36]. MAGNET is a multi-regional, multi-sectoral, applied general equilibrium model based on neo-classical microeconomic theory [37]. Two versions of this model were used for this paper. The first version was a standard model using the dynamic steering system to compile a GHG emissions model and associated databases. Adjustments were then made to this initial model so that the primary agricultural sector was excluded from carbon taxes in a second model version, i.e., the carbon tax was removed from equations modelling primary agricultural taxes within the model. The same baseline scenario was run on both model versions projecting the GTAP database over four time periods (2011–2017, 2017–2020, 2020–2025, and 2025–2030). Carbon tax scenarios were then imposed as counterfactual simulations in the years 2020, 2025 and 2030 in both models, where the respective nominal Aglink-Cosimo carbon taxes were deflated to real 2011 USD. The resulting percentage changes in the price of energy (i.e., aggregated price change of crude oil, gas, coal), as well as changes in the price of chemicals (i.e., proxy for mineral fertilisers and pesticides) were transmitted to the agricultural economic model. Since a carbon tax on crude oil and pesticides cannot be directly imposed in Aglink-Cosimo, the energy and pesticide price changes are taken from the first version of MAGNET (i.e., including carbon taxes on all economic sectors). Conversely, a carbon tax is directly imposed in the Aglink-Cosimo model for fertilisers by taking the price change from the second MAGNET model version (i.e., excluding carbon taxes for primary agriculture). For a more detailed description of the MAGNET model and its use for the scenario analysis presented in this paper, please see the Supplementary Materials.

In a similar manner, the implementation of a carbon tax in Aglink-Cosimo does not capture any change in emission intensities per agricultural activity. Such a change will occur when cost-efficient mitigation technologies and management practices get adopted, as long as the carbon price exceeds their implementation costs. To capture this technological effect, the Common Agricultural Policy Regional Impact Analysis (CAPRI) model is employed, which is a partial equilibrium, large-scale economic, global multi-commodity, agricultural sector model [38]. The CAPRI model does not have the same detailed global agricultural coverage as the Aglink-Cosimo model, but is able to calculate global marginal abatement cost curves (MACC; [13]). Consequently, the three carbon tax scenarios are implemented in the CAPRI model to identify the mitigation potentials through increased adoption of technology by agricultural producers as carbon taxes change. The simulated emission mitigation in CAPRI was then decomposed into production, structural and mitigation technology effects, and the resulting changes in emission intensities were then transferred into the Aglink-Cosimo carbon tax scenarios to get a complete picture of the effects of a low carbon economy on the global agricultural sector. The methodological approach of the paper is illustrated in Figure 2.

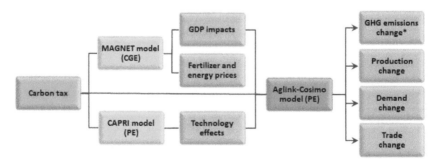

Figure 2. Methodological approach to estimating the effects of a low carbon economy on agriculture
Note: * The change in GHG emissions is decomposed in production, structural and technology effects.

4. Scenario Analysis and Discussion

4.1. Baseline

Baseline results show an increase in global agricultural non-CO_2 GHG emissions from 4.8 GtCO_2eq in 2016 to 5.4 GtCO_2eq by 2030 (Figure 3, left panel). This is in line with the projected FAOSTAT estimate of 5.8 Gt CO_2eq in 2030 [29] but well below other model (MAGNET, IMAGE) baseline projections of agricultural GHG emission [13]. In fact it is to be expected that the Aglink-Cosimo projected GHG emissions are lower than the ones estimated by the FAO, since the model does not include emissions from burning savannah and crop residues.

The projected increase in agricultural GHG emissions in the baseline is driven by increased demand for agricultural commodities by a growing population. However, due to yield gains per hectare of land and livestock head, global agricultural non-CO_2 emissions per capita are declining over time. Not surprisingly, commodities with the highest GHG intensities (kg CO_2eq/kg commodity) are cattle and sheep meat followed by chicken and pigmeat, whereas the lowest emission intensities are found in cereals, which is in line with Tubiello et al. [27]. This points the finger at animal husbandry as the main possible source of agricultural GHG mitigations. The Aglink-Cosimo baseline projects that about 76% of agricultural GHG emissions stem from animal husbandry, with enteric fermentation accounting for 50% of agricultural sector emissions. This falls in line with FAO estimates [29], which find that roughly 70% of agricultural GHG emissions stem from livestock.

Over the baseline period, average global meat (i.e., beef, pork, poultry, and sheep) consumption increases from 42.6 to 43.1 kg/capita/year in 2030, with notably poultry and sheep meat consumption increasing, while pork consumption declines and beef consumption per capita remains nearly unchanged. Even though the per capita consumption of bovine meat remains stable, an increasing population means increasing total demand for bovine meat, leading to an expansion of the global livestock inventories. The expected yield increases are not enough to meet the increased demand for food and feed, leading to an expansion of the global utilised agricultural area by 28 million hectares in 2030 compared to the base year 2016.

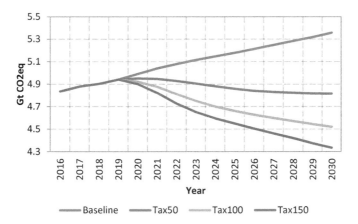

Figure 3. Global agricultural non-CO_2 GHG emissions in the baseline and tax scenarios. Source: own elaboration, Aglink-Cosimo model.

4.2. Macroeconomic Impacts on the Agricultural Sector

As shown in Figure 3, the introduction of carbon pricing reduces global agricultural non-CO_2 GHG emissions by 2030 between 10% (−0.540 GtCO_2eq) in Tax50 and 19% (−1.021 GtCO_2eq) in Tax150. Of these agricultural emission reductions, the macroeconomic spill-over effect from the rest of the economy of imposing the equivalent carbon taxes accounts for between 0.6% (−0.003 GtCO$_2$eq) in Tax50 and 1.3% (−0.013 GtCO$_2$eq) in Tax150. The largest macroeconomic impact stems from reductions in GDP (i.e., lower global income), followed by changes in input prices. The largest impacts on real GDP by 2030 occur in China (2.7%), India (2.6%) and Russia (2.5%) with a tax of 150 USD/t°CO$_2$eq. For a carbon tax of 50 USD/t (Tax50), crude oil, fertiliser, and pesticide prices are expected to, respectively, increase by 2.0–15%, 2.2–17% and 0.2–1.7% in the period 2020–2030. As the carbon tax increases to 150 USD/t CO$_2$eq (Tax150), the price of crude oil, fertiliser, and pesticides are projected to increase by 43, 48, and 4.5%, respectively, in 2030. The increase of input prices for Tax100 are in corresponding ranges between the price increases of Tax50 and Tax150.

The macroeconomic spill-over effects account for only as much as 1.3% of global agricultural GHG emissions reduction, but at the country level contributions might vary significantly. For example, in Canada, the United States and China, macroeconomic spill-over effects of a 150 USD/t CO$_2$eq tax account for 5.8, 4.6 and 3.3%, respectively, of total mitigation in their domestic agricultural sectors. Other regions and countries with less intensive agricultural production (i.e., relying less on the input use of fertiliser and fuel) are less affected by the macroeconomic changes induced by moving to a global low carbon economy. For instance, the macroeconomic impacts in the three scenarios actually lead to an increase in the agricultural emissions of countries like Argentina and Australia.

4.3. Decomposition of Mitigation and Production Impacts

The introduction of carbon taxes in the agricultural sector could reduce global agricultural non-CO$_2$ GHG emissions by 2030 between 10% in Tax50 (0.540 Gt CO$_2$eq) and 19% in Tax150 (1.021 Gt CO$_2$eq). These declines, particularly those observed in Tax150, are fairly consistent with other studies aiming to estimate the efforts required to achieve the goal of the 2 °C scenario by 2030, such as the 20% (1.07 Gt CO$_2$eq) reduction of agricultural GHG emissions estimated by Kitous et al. [2] and the 11–18% (0.92–1.37 GtCO$_2$eq) identified by Wollenberg et al. [4]. Given our medium-term time horizon of implementing a carbon tax (2020–2030), and the phasing in of taxes over this period, a larger emission reduction is not likely. In other studies, typically with a longer time span, a 20-30% emission reduction is feasible. Frank et al. [13] compare the mitigation potential of imposing a homogenous carbon tax on the agricultural sector across four models, highlighting that the models already at a carbon price of 100 USD/t CO$_2$eq show a significant potential for emission reductions of 1.6 to 2.6 Gt CO$_2$eq/year by 2050, which is equivalent to a 20–35% reduction compared to the baseline (it should be noted that this study assumes a 20 year longer time span (2030–2050) and the carbon tax is in real USD, which in nominal prices would be equivalent to a tax of 200-250 USD in 2050, depending on the rate of inflation, when implemented in Aglink-Cosimo, which uses nominal USD). In the simulation exercise of Frank et al. [13], technical and structural changes account for 75–80% of agricultural emissions reduction, however, this is over a longer time horizon (100 USD/t CO$_2$eq in 2050). The remaining mitigation of 20–25% is achieved through a reduction in domestic production. The decomposition of our scenario results are in line with the findings of Frank et al. [13], with the major mitigation contributions coming from technology and management options (74% by 2030 in Tax150). Structural adjustments contribute 16% to overall mitigation, i.e., changes in trade and production mix (e.g., switching from ruminant to non-ruminant products), whereas production reductions only contribute 10%. In this decomposition exercise countries and world production of commodities were aggregated into two categories—animal-based and crop-based production—using their relative dry matter content, as well as aggregating their carbon emissions. This enabled us to capture structural changes due to changes in yields and the dynamic effects of relative movement of production between sectors with varying carbon emission intensities, within the aggregated sectors.

The mitigation development presented in Figure 4 shows that mitigation from applying new technologies becomes more attractive as carbon prices increase. At low carbon prices, the cost-efficient mitigation approach is through commodity trade or shifts in consumption preferences for agricultural products, i.e., structural effects. Adoption of new technologies is often not attractive due to high implementation costs. However, as carbon taxes increase, technology application becomes more attractive and cost-efficient. The underlying logic is that the introduction of a carbon tax increases farmer's incentives to adopt new mitigation technologies and management practices, as well as increases consumer's incentives to change their consumption habits. The mitigation measures will be adopted as long as the carbon price exceeds the costs per tonne CO_2eq saving of a mitigation option. Accordingly, as the carbon price increases, the share of the technological mitigation effect increases compared to the decreasing share of structural change mitigation effect (as shown in Figure 4). However, Frank et al. [13] show that there are limits to the emission mitigation that can be achieved by technology adoption and structural changes. Consequently, their mitigation share decreases the further carbon prices rise beyond the tax level in our analysis, and mitigation has to be achieved by reduction in agricultural production levels [13]. It has to be noted that the technology effect in our results is based on the work by Lucas et al. [34] and calculated by the CAPRI model (www.capri-model.org), which tends to have relatively small production effects (and hence higher technology and structural effects) compared to other models at high carbon prices. This is mainly because CAPRI has a detailed cross-price matrix, which allows for strong substitution between products (for example ruminant and non-ruminant products), leading to a stabilisation of consumption, and hence, production.

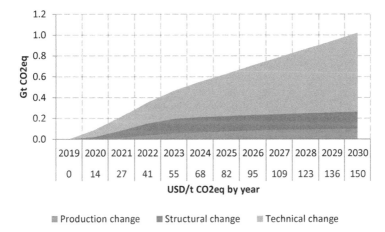

Figure 4. Decomposition of mitigation effects for the period 2020–2030 for the carbon price trajectory underlying the scenario Tax150. Source: own elaboration, Aglink-Cosimo model.

The decomposition of global mitigation mechanisms presented in Figure 4 is not always reflected at the individual country level (Table 1). For the understanding of country level results it is important to keep in mind that at the global level, trade falls under the structural effects. However, at the national (country) level, part of the global trade effect becomes a national production effect. For instance, total agricultural production in Brazil is decreasing, contributing to 26% of the total agricultural emissions mitigation (most of the adjustments in the beef sector). The European Union is characterized by a relatively emission-efficient agricultural sector (i.e., relatively low emissions per kg of commodity produced) compared to other countries. Accordingly, the simulated carbon taxes increase the European Union´s competitiveness, which leads to an increase in domestic production, and hence also to a 2% increase in the related emissions. LDCs are exempt from the simulated carbon tax in our scenarios. Accordingly, their production and related emissions are expected to increase, while at the same time they have no incentives to increase the adoption in mitigation technologies (Table 1).

Table 1. Decomposition of mitigation effects in 2030 for selected countries, Scenario Tax150.

	Mitigation (Gigatonnes of Carbon Dioxide Equivalents)	%-Share of Mitigation from Changes in		
		Production Levels	Technical Options	Structural Adjustments
Brazil	0.230	26	36	39
China	0.185	10	66	23
United States	0.069	3	72	25
Pakistan	0.054	15	77	8
European Union	0.053	−2	98	4
Indonesia	0.051	8	84	8
Least Developed Countries	−0.035			76

Source: own elaboration, Aglink-Cosimo model. Note: a negative value implies an increase in emissions.

Looking closer at commodity and country level results, the impact of the carbon taxes are, not surprisingly, most pronounced in the livestock sector. Over the baseline period, agricultural sector emissions increase by 0.52 $GtCO_2eq$, with the majority stemming from increased production of dairy products and bovine and sheep meat production. The introduction of a carbon tax mainly reduces the production of bovine meat and milk production (Figure 5, left panel), reducing total emissions by 0.50 $GtCO_2eq$ in Tax150 by 2030, which represents a total reduction of 1.02 $GtCO_2eq$ (Figure 5, right panel).

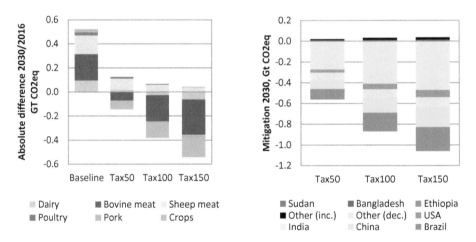

Figure 5. Changes in agricultural non-CO_2 GHG emissions by commodity (2030 compared to 2016, left panel) and by region (scenarios compared to baseline in 2030, right panel). Source: own elaboration, Aglink-Cosimo model. Note: other (dec.) refers to "other countries with decreasing emissions", and other (inc.) refers to "other countries with increasing emissions".

The changes in domestic production are driven by the relative impact of the carbon tax, where countries with relatively low GHG intensities (i.e., kg CO_2eq/kg commodity) become more competitive on the world market and global exports increase (cheese 6% and butter 8% in Tax 150 by 2030). The carbon tax, however, also reallocates production between countries or regions. Notably, the production of milk and bovine meat increases in the United States and the European Union (Figure 6). At the same time, India increases its beef production and New Zealand sees its competitive dairy sector benefiting. At the other end, Brazil, China and Argentina reduce their beef production and India, Brazil, Pakistan have their milk production negatively affected.

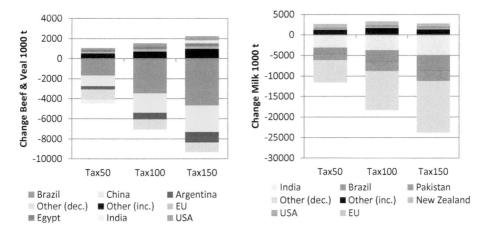

Figure 6. Changes in beef (left panel) and milk production (right panel) in 2030 compared to the baseline. Source: own elaboration, Aglink-Cosimo model. Note: other (dec.) refers to "other countries with decreasing emissions", and other (inc.) refers to "other countries with increasing emissions".

Global per capita consumption of bovine meat (−9%) butter (−3%) and fresh dairy products (−2%) declines (Figure 7) as domestic consumer prices increase. In the United States, for example, consumers reduce their average bovine meat consumption by 3%, consuming 2% more poultry and pork meat instead. Nonetheless, US bovine meat production increases by 3%, as exports expand by 44%, driven by the relatively more competitive production industry in the United States.

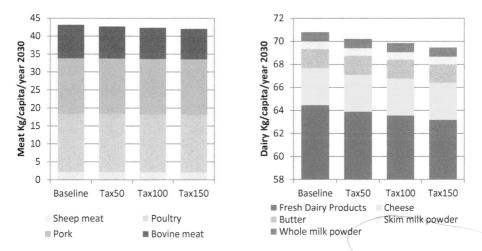

Figure 7. Global per capita consumption of meat (left panel) and dairy (right panel) in 2030. Source: own elaboration, Aglink-Cosimo model.

4.4. Impacts on Bovine Meat Production and Consumption Patterns

The impact of the carbon tax on production and consumption patterns is worth further examination, as a change in consumer diets is seen as a possibility to reduce the carbon footprint of the agricultural sector, especially if the diet favours eating less bovine meat. Switching from ruminant to non-ruminant meat consumption would reduce CO_2eq emissions by 85% per kilocalorie [39]. Such a movement in consumer's global preferences can be expected to have a large impact in countries where per capita consumption is high (e.g., the United States). In the baseline, 49% of the global bovine meat production is consumed in the United States, China, Brazil and the European Union (Table 2). By 2030, per capita consumption of bovine meat is projected to decline in these countries, with the exception of China, where per capita consumption of meat increases from 5.7 to 7.1 kg, which is still well below the projected consumption levels of the United States, Brazil and the European Union (35.8, 37.5, 14.5 kg per capita/year, respectively).

Table 2. Bovine meat consumption, production, imports and exports, 2030 baseline.

	Consumption	Production	Exports	Imports
	Relative share in global (%)			
United States	16.2	15.6	11.3	14.9
China	12.7	11.4	0.3	8.7
Brazil	10.9	14.3	20.4	0.4
European Union	9.4	9.4	2.5	2.3
Australia	1.1	3.9	17.3	0.1
India	1.4	4.1	15.8	0.0
Vietnam	2.0	0.5	0.0	9.9
Total	53.8	59.1	67.7	36.3
	Total global (1000 t)			
World	78623	78747	13167	13032

Source: own elaboration, Aglink-Cosimo model.

The introduction of a 150 USD/t CO_2eq carbon tax and the resulting commodity price increase give an incentive to consumers to change their consumption habits to less emission-intensive products. In the United States, the farm gate price of bovine meat is projected to increase by 32%, while the consumer price increases by 10% (i.e., the initial farm gate price is roughly one-third of the consumer price). This increase in consumer prices leads to a reduction of bovine meat consumption in the United States by 3.2%, while pork and poultry meat consumption increases by 2.0%. In countries such as Brazil, where the farm gate price constitutes a larger share of the wholesale consumer price, the price signals are stronger and bovine meat consumption declines by 18%, with consumers switching their consumption habits to poultry and pork meat. This means that price signals in highly developed countries are weaker and changes in diets will perhaps have to come from changing preferences, for example increasing demand for meat substitutes driven by vegetarians and flexitarians. The introduction of the 150 USD/t CO_2eq carbon tax only reduces bovine meat consumption by 9% (Table 3), so that further changes in diets could be required to meet the 2 °C scenarios of the Paris Agreement by 2050 (Figure 1).

Table 3. Bovine meat production in 2030.

	Base	Tax50	Tax100	Tax150
	1000 t	% change compared to base		
Africa	7956	4	6	8
China	8939	−12	−22	−30
Asia	9813	−4	−1	2
Brazil	11,211	−15	−31	−42
Latin America	9961	−7	−9	−11
European Union	7410	2	3	3
Europa	2804	−4	−2	1
North America	13,768	1	2	3
Oceania	3641	−2	−4	−6
World	78,747	−4	−7	−9

Source: own elaboration, Aglink-Cosimo model. Note: Asia: excluding China; Europe: excluding the European Union; Latin America: Middle and South America (excluding Brazil); North America: United States and Canada.

4.5. Impacts on Land Use

In the baseline, total utilised agricultural land (UAA) increases by 28 million hectares over the projection period, which is an average yearly increase of 0.04% (Figure 8, left panel). This does not consider any land use change from forestry and other land use into cropland or grassland, or the associated GHG emissions from net forest conversion (in the model used, LULUCF-related CO_2 emissions and sinks form agricultural production are not accounted for; however, the model projects changes in land use, which already can give an indication on the related CO_2 emissions). Nonetheless, scenario results suggest that imposing a carbon tax (Tax150) actually increases global UAA by an additional 30 million hectares, with pasture land expanding by 33 million hectares (i.e., 3 million ha of cropland is converted to pasture land). The main bulk of this increased pasture land is found in Africa, where bovine meat production increases by 8%.

Increasing UAA generally comprises CO_2 emissions, whereas removing land from agricultural production and converting it to perennial plants, such as trees, grass or shrubs, is generally regarded as a positive contribution to climate change mitigation through carbon sequestration [40]. However, the net effect of the global increase or decrease in UAA on CO_2 emissions and removals is not straightforward. For example, the reduction in UAA in the Tax150 scenario does not necessarily lead to a net decrease in global LUC-related CO_2 emissions, as soil carbon emissions and removals depend on many factors, such as the management system and location [40–42]. Moreover, the decrease or increase in UAA is not universal in the scenarios, and while many countries reduce their agricultural land in the Tax150, the change in UAA in 2030 also comprises countries expanding their UAA, among others the African countries, namely Angola and Nigeria (Figure 8, right panel). This is not surprising, since these two countries are classified as LDCs, and therefore, are exempted from the simulated carbon tax. This makes their domestic agricultural production more competitive, which leads to production increases that also involve increasing UAA. Given the regional distribution of land moving in and out of agricultural production, it is difficult to quantify how large the net contribution of the change in land use to global GHG emissions mitigation is. Taking pasture land out of production in the outback of Australia compared to pasture land in the European Union does not imply the same reduction of GHG emissions per hectare (see the discussions and literature in Powlson et al. [40] and Oertel et al. [41]). Further work needs to be done to account for LULUCF emissions and get a clearer picture of the (potential) contribution of the agricultural, forestry and other land use sectors to GHG mitigation.

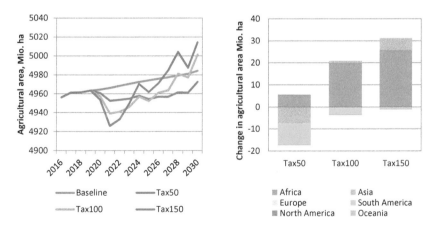

Figure 8. Development in global utilised agricultural area 2016–2030 (left panel) and its regional change in the scenarios compared to the baseline in 2030 (right panel). Source: own elaboration, Aglink-Cosimo model.

5. Conclusions

Limiting climate change to ensure global temperature increases remain 2 °C below pre-industrial levels by the end of the century requires substantial reductions of GHG emissions and the transition to a climate-friendly, low carbon economy. A transition to a lower carbon intensive economy has large implications from both regional and global perspectives. Moreover, it needs to consider not only the environmental dimension but also the economic and societal ones. Policies aiming at a decarbonized economy can have important collateral effects in terms of people's discontentment, as, for example, recent movements in France have shown [43,44]. Furthermore, the increase of prices linked to discretionary climate change mitigation policies can have negative effects on poor economies, and could increase food insecurity [12] and migration flows [45]. These elements highlight how the necessary transition to a low carbon economy must be carefully designed. Accordingly, the implementation of GHG mitigation policies in a specific sector needs to be "fair" in the sense of not only taking into account the long-term benefits (i.e., the GHG mitigation goal and limiting climate change) but also short and medium-term costs (transition), and it should be global, such as to minimise emission leakage and effectively reduce GHG emissions [46].

Using the Paris Agreement as a framework for limiting global temperature rises, in this paper an empirical study is performed on how policies aiming at a global lower carbon intensive economy could be transmitted into agricultural markets. For the analysis, an updated version of the Aglink-Cosimo model is employed to simulate three carbon tax scenarios, specifically accounting for the macroeconomic and technological effects inherent in moving to a global low carbon economy (captured with the MAGNET and CAPRI models). Within this scenario design, homogenous taxes on agricultural non-CO_2 emissions (i.e., methane and nitrous oxide) are implemented globally, with the exception of least developed countries, and increased progressively to 50 USD per tonne of CO_2eq, 100 USD/t CO_2eq, and 150 USD/t CO_2eq, respectively, by 2030. Simulation results show that global GHG emissions from the primary agricultural sector are reduced by between 10% and 19% in 2030 compared to the baseline.

The analysis indicates that for the net mitigation of global agricultural GHG emissions, it specifically matters where (i.e., in which country or region) production is affected by climate change mitigation policies. Larger (lower) effects are expected in countries (and commodities) with relatively high (low) emissions per production unit. The results highlight the importance of GHG emission reduction policies on agricultural markets over a medium-term time horizon, as the sector is affected by both direct (i.e., through emission abatement commitments within the agricultural sector) and indirect (i.e., through changes in prices for fossil fuel intensive goods and macroeconomic variables) mitigation policies. For instance, it is shown how emission reductions compatible with the Paris Agreement can have significant effects on agricultural production, especially when looking at the regional impacts. These results also underline the importance of taking climate-change-related policies into account when producing agricultural market outlooks. In this respect, enabling Aglink-Cosimo to account for agricultural emissions and respective mitigation efforts is an essential development, especially considering that the model is used by the OECD and FAO to produce agricultural market projections that establish the benchmark for many other agricultural economic models. However, for future research the Aglink-Cosimo model needs to be further developed, for example to include the adoption of new GHG emission abatement technologies and the contribution of structural change within farming. Moreover, the model should be modified to account for CO_2 emissions and removals related to land use and land use changes, to get a broader picture of the possible contribution (and resulting impacts) of the agricultural sector to a global low carbon economy. These aspects are important for the future assessment of both the full potential of the agricultural sector to contribute to achieving the goal of the Paris Agreement, as well as the related impacts to agricultural market developments and potential effects on food security.

Our results show that the technological development induced by the carbon tax can substantially help mitigate GHG emissions, and hence the need to reduce agricultural production levels globally. Technological development is especially important in some developing countries that would be relatively more affected by global carbon taxes, as they are usually characterised by higher emission intensities (kg CO_2eq/kg commodity) and are less competitive on the global agricultural commodity markets. This points towards the importance of both (i) technology change and transfer, to reduce emission intensities especially in developing countries (i.e., need to modernize agricultural production systems), and (ii) more sophisticated and differentiated policy approaches for the agricultural sector, to achieve a significant contribution towards the move to a global low carbon economy.

Supplementary Materials: The following are available online at http://www.mdpi.com/2071-1050/11/8/2349/s1.

Author Contributions: Conceptualization, I.P.D. and H.J.; Methodology, I.P.D., H.J., P.L., J.H. and G.P.; Formal Analysis, H.J., T.F., I.P.D. and P.L.; Writing-Original Draft Preparation, H.J., T.F. and I.P.D.; Writing-Supplementary Material, G.P. and H.J.; Project Administration, I.P.D.

Funding: This research received no external funding.

Acknowledgments: This material is based upon work supported primarily by the Joint Research Centre of the European Commission under the specific contract 154208.X37. We would like to thank Pierre Charlebois for his contribution to improving the Aglink-Cosimo model, facilitating the analysis of the impacts of a less carbon intensive economy on the agricultural sector.

Conflicts of Interest: The authors declare no conflict of interest.

References

1. UNFCCC. Adoption of the Paris Agreement. United Nations Framework Convention on Climate Change, FCCC/CP/2015/L.9/Rev.1. 2015. Available online: http://unfccc.int/resource/docs/2015//l09r01.pdf (accessed on 8 May 2018).

2. Kitous, A.; Keramidas, K.; Vandyck, T.; Saveyn, B. *GECO 2016—Global Energy and Climate Outlook, Road from Paris*; JRC Science for Policy Reports, EUR 27952 EN; Publications Office of the European Union: Luxembourg, 2016.

3. Kitous, A.; Keramidas, K. *GECO 2016—GHG and Energy Balances*; JRC Technical Reports, EUR 27976 EN; European Commission: Brussels, Belgium, 2016.

4. Wollenberg, E.; Richards, M.; Smith, P.; Havlík, P.; Obersteiner, M.; Tubiello, F.N.; Campbell, B.M. Reducing emissions from agriculture to meet the 2 °C target. *Glob. Chang. Biol.* **2016**, *22*, 3859–3864. [CrossRef] [PubMed]

5. Reisinger, A.; Havlik, P.; Riahi, K.; van Vliet, O.; Obersteiner, M.; Herrero, M. Implications of alternative metrics for global mitigation costs and greenhouse gas emissions from agriculture. *Clim. Chang.* **2013**, *117*, 677–690. [CrossRef]

6. Gernaat, D.E.H.J.; Calvin, K.; Lucas, P.L.; Luderer, G.; Otto, S.A.C.; Rao, S.; van Vuuren, D.P. Understanding the contribution of non-carbon dioxide gases in deep mitigation scenarios. *Glob. Environ. Chang.* **2015**, *33*, 142–153. [CrossRef]

7. Rogelj, J.; Popp, A.; Calvin, K.V.; Luderer, G.; Emmerling, J.; Gernaat, D.; Krey, V. Scenarios towards limiting global mean temperature increase below 1.5 °C. *Nat. Clim. Chang.* **2018**, *8*, 325. [CrossRef]

8. Obersteiner, M.; Bednar, J.; Wagner, F.; Gasser, T.; Ciais, P.; Forsell, N.; Peñuelas, J. How to spend a dwindling greenhouse gas budget. *Nat. Clim. Chang.* **2018**, *8*, 7. [CrossRef]

9. Van Vuuren, D.P.; Stehfest, E.; Gernaat, D.E.; Berg, M.; Bijl, D.L.; Boer, H.S.; Hof, A.F. Alternative pathways to the 1.5 °C target reduce the need for negative emission technologies. *Nat. Clim. Chang.* **2018**, *8*, 391. [CrossRef]

10. Doelman, J.C.; Stehfest, E.; Tabeau, A.; van Meijl, H.; Lassaletta, L.; Gernaat, D.E.; van der Sluis, S. Exploring SSP land-use dynamics using the IMAGE model: Regional and gridded scenarios of land-use change and land-based climate change mitigation. *Glob. Environ. Chang.* **2018**, *48*, 119–135. [CrossRef]

11. Van Meijl, H.; Havlik, P.; Lotze-Campen, H.; Stehfest, E.; Witzke, P.; Domínguez, I.P.; Humpenöder, F. Comparing impacts of climate change and mitigation on global agriculture by 2050. *Environ. Res. Lett.* **2018**, *13*, 064021. [CrossRef]

12. Hasegawa, T.; Fujimori, S.; Havlík, P.; Valin, H.; Bodirsky, B.L.; Doelman, J.C.; Mason-D'Croz, D. Risk of increased food insecurity under stringent global climate change mitigation policy. *Nat. Clim. Chang.* **2018**, *8*, 699. [CrossRef]

13. Frank, S.; Havlík, P.; Stehfest, E.; van Meijl, H.; Witzke, P.; Pérez-Domínguez, I.; Tabeau, A. Agricultural non-CO$_2$ emission reduction potential in the context of the 1.5 °C target. *Nat. Clim. Chang.* **2019**, *9*, 66. [CrossRef]

14. OECD-FAO. Aglink-Cosimo Model Documentation. A Partial Equilibrium Model of the World Agricultural Markets. 2015. Available online: http://www.agri-outlook.org/abouttheoutlook/Aglink-Cosimo-model-documentation-2015.pdf (accessed on 8 May 2018).

15. Araujo Enciso, S.R.; Pérez-Domínguez, I.; Santini, F.; Helaine, S. *Documentation of the European Commission's EU Module of the Aglink-Cosimo Modelling System*; JRC Science and Policy Reports; Publications Office of the European Union: Luxembourg, 2015; Available online: http://publications.jrc.ec.europa.eu/repository/bitstream/JRC92618/jrc92618%20online.pdf (accessed on 18 December 2017).

16. OECD-FAO. *OECD-FAO Agricultural Outlook 2017–2030*; Organisation for Economic Cooperation and Development, Paris, and Food and Agricultural Organisation of the United Nations, OECD Publishing and FAO: Rome, Italy, 2017.

17. European Commission. *EU Agricultural Outlook. For the Agricultural Markets and Income 2017–2030*; DG Agriculture and Rural Development, European Commission: Brussels, Belgium, 2017; Available online: https://ec.europa.eu/agriculture/sites/agriculture/files/markets-and-prices/medium-term-outlook/2017/2017-fullrep_en.pdf (accessed on 8 May 2018).

18. Frank, S.; Beach, R.; Havlík, P.; Valin, H.; Herrero, M.; Mosnier, A.; Obersteiner, M. Structural change as a key component for agricultural non-CO$_2$ mitigation efforts. *Nat. Commun.* **2018**, *9*, 1060. [CrossRef]

19. Nekhay, O.; Fellmann, T.; Gay, S.H. A free trade agreement between Ukraine and the European Union: potential effects on agricultural markets and farmers' revenues. *Post-Communist Econ.* **2012**, *24*, 351–363. [CrossRef]

20. Fellmann, T.; Hélaine, S.; Nekhay, O. Harvest failures, temporary export restrictions and global food security: The example of limited grain exports from Russia, Ukraine and Kazakhstan. *Food Secur.* **2014**, *6*, 727–742. [CrossRef]

21. Enciso, S.R.A.; Fellmann, T.; Dominguez, I.P.; Santini, F. Abolishing biofuel policies: Possible impacts on agricultural price levels, price variability and global food security. *Food Policy* **2014**, *61*, 9–26. [CrossRef]

22. Kavallari, A.; Fellmann, T.; Gay, S.H. Shocks in economic growth = shocking effects for food security? *Food Secur.* **2014**, *6*, 567–583. [CrossRef]

23. Santini, F.; Ronzon, T.; Dominguez, I.P.; Enciso, S.R.A.; Proietti, I. What if meat consumption would decrease more than expected in the high-income countries? *Bio-Based Appl. Econ.* **2017**, *6*, 37–56.

24. Charlebois, P.; Kanadani Campos, S.; Pérez Domínguez, I.; Jensen, H. *Enhancing the Brazilian Land Use Module in Aglink-Cosimo*; EUR 28633 EN; European Commission: Brussels, Belgium, 2017.

25. Thompson, W.; Dewbre, J.; Westhoff, P.; Schroeder, K.; Pieralli, S.; Pérez-Domínguez, I. *Introducing Medium- and Long-Term Productivity Responses in Aglink-Cosimo*; JRC Technical Reports, EUR 28560 EN; European Commission: Brussels, Belgium, 2017.

26. IPCC. *2006 IPCC Guidelines for National Greenhouse Gas Inventories*; Prepared by the National Greenhouse Gas Inventories Programme; IGES: Tokyo, Japan, 2006.

27. Tubiello, F.N.; Salvatore, M.; Cóndor Golec, R.D.; Ferrara, A.; Rossi, S.; Biancalani, R.; Flammini, A. *Agriculture, Forestry and Other Land Use Emissions by Sources and Removals by Sinks*; Statistics Division, Food and Agriculture Organization: Rome, Italy, 2014.

28. Tubiello, F.N.; Cóndor-Golec, R.D.; Salvatore, M.; Piersante, A.; Federici, S.; Ferrara, A.; Jacobs, H. *Estimating Greenhouse Gas Emissions in Agriculture: A Manual to Address Data Requirements for Developing Countries*; FAO: Roma, Italy, 2015.

29. FAOSTAT. *FAOSTAT Emissions Database*; Food and Agriculture Organization of the United Nations (FAO): Roma, Italy, 2018; Available online: http://www.fao.org/faostat/en/#data/GT (accessed on 8 May 2018).

30. Jansson, T.; Pérez Domínguez, I.; Weiss, F. Estimation of greenhouse gas coefficients per commodity and world region to capture emission leakage in European agriculture. Paper Presented at the 119th EAAE Seminar "Sustainability in the Food Sector", Capri, Italy, 30 June–2 July 2010.

31. Pérez Domínguez, I.; Fellmann, T.; Weiss, F.; Witzke, P.; Barreiro-Hurle, J.; Himics, M.; Jansson, T.; Salputra, G.; Leip, A. *An Economic Assessment of GHG Mitigation Policy Options for EU Agriculture (EcAMPA 2)*; JRC Science for Policy Report; European Commission, Publications Office of the European Union: Luxembourg, 2016.

32. Pérez Domínguez, I.; Fellmann, T.; Witzke, H.P.; Jansson, T.; Oudendag, D.; Gocht, A.; Verhoog, D. *Agricultural GHG emissions in the EU: An Exploratory Economic Assessment of Mitigation Policy Options*; JRC Scientific and Policy Reports; European Commission: Seville, Spain, 2012.

33. Van Doorslaer, B.; Witzke, P.; Huck, I.; Weiss, F.; Fellmann, T.; Salputra, G.; Leip, A. *An Economic Assessment of GHG Mitigation Policy Options for EU Agriculture (EcAMPA). JRC Science for Policy Report*; European Commission, Publications Office of the European Union: Luxembourg, 2015.

34. Lucas, P.L.; van Vuuren, D.P.; Olivier, J.G.; Den Elzen, M.G. Long-term reduction potential of non-CO_2 greenhouse gases. *Environ. Sci. Policy* **2007**, *10*, 85–103. [CrossRef]

35. Lamb, A.; Green, R.; Bateman, I.; Broadmeadow, M.; Bruce, T.; Burney, J.; Goulding, K. The potential for land sparing to offset greenhouse gas emissions from agriculture. *Nat. Clim. Chang.* **2016**, *6*, 488. [CrossRef]

36. Aguiar, A.; Narayanan, B.; McDougall, R. An overview of the GTAP 9 data base. *J. Glob. Econ. Anal.* **2016**, *1*, 181–208. [CrossRef]

37. Kavallari, A.; van Meijl, H.; Powell, J.; Rutten, M.; Shutes, L.; Tabeau, A. *The MAGNET Model: Module Description*; LEI Report 14-057; LEI Wageningen UR (University & Research Centre): Wageningen, The Netherlands, 2014.

38. Britz, W.; Witzke, P. CAPRI Model Documentation 2014. University of Bonn. Institute for Food and Resource Economics, 2014. Available online: http://www.caprimodel.org/docs/capri_documentation.pdf (accessed on 8 May 2018).

39. Girod, B.; van Vuuren, D.P.; Hertwich, E.G. Climate policy through changing consumption choices: Options and obstacles for reducing greenhouse gas emissions. *Glob. Environ. Chang.* **2014**, *25*, 5–15. [CrossRef]

40. Powlson, D.S.; Whitmore, A.P.; Goulding, K.W. Soil carbon sequestration to mitigate climate change: A critical re-examination to identify the true and the false. *Eur. J. Soil Sci.* **2011**, *62*, 42–55. [CrossRef]

41. Oertel, C.; Matschullat, J.; Zurba, K.; Zimmermann, F.; Erasmi, S. Greenhouse gas emissions from soils—A review. *Chem. Der Erde-Geochem.* **2016**, *76*, 327–352. [CrossRef]

42. Thamo, T.; Pannell, D.J.; Kragt, M.E.; Robertson, M.J.; Polyakov, M. Dynamics and the economics of carbon sequestration: common oversights and their implications. *Mitig. Adapt. Strateg. Glob. Chang.* **2017**, *22*, 1095–1111. [CrossRef]

43. Klenert, D.; Mattauch, L.; Combet, E.; Edenhofer, O.; Hepburn, C.; Rafaty, R.; Stern, N. Making carbon pricing work for citizens. *Nat. Clim. Chang.* **2018**, *8*, 669–677. [CrossRef]

44. Nature. Wanted: A fair carbon tax. Nature Editorial. *Nature* **2018**, *564*, 161.

45. Black, R.; Kniveton, D.; Schmidt-Verkerk, K. Migration and climate change: towards an integrated assessment of sensitivity. *Environ. Plan. A* **2011**, *43*, 431–450. [CrossRef]

46. Domínguez, I.P.; Fellmann, T. The need for comprehensive climate change mitigation policies in European agriculture. *EuroChoices* **2015**, *14*, 11–16. [CrossRef]

Article

Impact of Artificially Simulated Precipitation Patterns Change on the Growth and Morphology of *Reaumuria soongarica* Seedlings in Hexi Corridor of China

Yanfei Xie, Yi Li *, Tingting Xie, Ruiling Meng and Zhiqiang Zhao

College of Forestry, Gansu Agricultural University, Lanzhou 730070, China; xieyf@gsau.edu.cn (Y.X.); xieting1026@126.com (T.X.); Mrling@tom.com (R.M.); zhaozq1996@163.com (Z.Z.)
*   Correspondence: liyi@gsau.edu.cn

Received: 11 February 2020; Accepted: 18 March 2020; Published: 20 March 2020

Abstract: Climate change has altered the existing pattern of precipitation and has an important impact on the resistance and adaptability of desert plants. However, the interactive impact and the main characteristics of changes in precipitation amount and precipitation frequency on desert plants are unclear. *Reaumuria soongarica* seedlings were treated by artificially simulating changes in precipitation (30% reduction and 30% increase) and its frequency (50% reduction). We first introduced three morphological indicators (i.e., main root length/plant height ratio (*RHR*), above-ground radial density (*ARD*), and below−ground radial density (*BRD*)) and drew an abstract figure of seedling growth. This experiment confirmed the following: (1) The increase in precipitation noticeably increased the plant height, above-ground biomass, and total biomass of seedlings. (2) The plant height and the biomass of seedlings were more affected by precipitation amount than by precipitation frequency. No interaction was found between precipitation amount and precipitation frequency on the growth of seedlings. (3) The response of *RHR* to precipitation changes was extremely significant, increasing with decreasing precipitation and vice versa. (4) The *ARD* first increased then remained constant as precipitation increased, while *ARD* first decreased and then increased with decreasing precipitation. When precipitation increases, the *BRD* increases and the root system becomes relatively thicker and shorter, and vice versa. In this regard, *R. soongarica* seedlings mainly adapt to their resource supply by adjusting plant height, root length, thickness and biomass.

Keywords: climate change; desert ecosystems; precipitation patterns; growth and morphology; *Reaumuria soongarica*

---

1. Introduction

Climate change is altering existing precipitation patterns [1,2]. The time and intensity of precipitation may change [3]. In the arid and semi-arid regions of northwestern China, precipitation will increase by 30 to 100 mm in the next 100 years [2]. At the same time, it will be accompanied by a trend of increasing precipitation intervals, decreasing small precipitation events and increasing extreme precipitation events [2]. Globally, the proportion of land surface under extreme drought is predicted to increase from 1%–3% at present to 30% by the end of 2090 [4]. The change of existing pattern of precipitation affects the amount and distribution of many physical, chemical, and biological factors inducing an alteration of available resources and a consequent destabilization of the ecological successions [5]. However, the interactive impact and the main characteristics of changes in precipitation amount and precipitation frequency to desert ecological are unclear. Due to their morphological and functional characteristics, plants show their conservation or alteration status in every moment.

Therefore, plants can be considered general biological indicators of the environment where they develop, especially in largescale plant communities [5]. This makes it possible that artificial simulated precipitation patterns change, as the anthropogenic disturbance was used to predict the possible impact of precipitation pattern changes in the future on desert plants and can be observed at different scales and grains of definition.

It is generally believed that the increase in precipitation will increase the above-ground biomass of desert shrubs [6–8] and decrease that of herbaceous plants [9], a decrease in precipitation increases the underground biomass of desert shrubs [6], which may be related to the lifestyle of the plants. However, other studies have also shown that the below−ground biomass of shrubs seedlings decreases with increased precipitation [8], and the biomass of herbaceous plants increases with increased precipitation [10], precipitation increases from 95 to 283 mm, which may be differently sensitive to ecosystem disturbance in different growth stages [11,12]. Some researchers have determined that the frequency of precipitation is also a key factor affecting plant growth [11,13,14]. The main change in precipitation will likely be in the intensity, frequency, and duration of events, but these characteristics are seldomly analyzed in observations or models [15]. Moreover, many scholars only study the impact of precipitation, but few compare the frequency and interactive impacts on individual plant growth by multiple dimensions of precipitation pattern changes (such as precipitation amount and precipitation frequency).

Reaumuria soongarica is an ultra-drought shrub of the Tamaricaceae family. It is the main dominant species and founder species of plant communities in arid and semi-arid desert areas. It has a large distribution area, strong stress resistance, barren resistance, drought resistance, and strong sand collection capacity. It is widely distributed in Central Asia, Western Asia, Southern Europe, and North Africa and is mainly distributed in the northwest region in China. It plays an important role in maintaining the ecological stability of the desert. Seeding is a young plant that is grown from a seed, rather than from a cutting or bulb, for example [16]. The seedling stage of a plant is the core of plant population renewal [17]. It also has a more fragile development stage during plant growth and it is extremely sensitive to ecosystem disturbance [11,12]. At present, research on the impact of precipitation and changes in precipitation frequency on the morphology of *R. soongarica* seedlings has been reported [18–20], but has only studied annual *R. soongarica* seedlings. The complete life cycle of *R. soongarica* is several decades, so it is necessary to study the seedlings for a longer time. We take *R. soongarica* seedlings as the research object and analyze the deviations in their morphological characteristics by artificially controlling the precipitation amount (increase or decrease by 30%) and precipitation frequency (F) (decrease by 50%), in order to reveal the response and adaptation strategies of *R. soongarica* seedlings to precipitation pattern changes. We aim to answer the following questions: (1) What is the impact of precipitation pattern changes on the morphological characteristics of *R. soongarica* seedlings? (2) Can increasing precipitation and decreasing precipitation frequency promote the growth of *R. soongarica* seedlings? (3) What is the adaptive strategy of *R. soongarica* seedlings to precipitation pattern changes?

2. Materials and Methods

2.1. Study Area

The research area was selected at the National Field Observation and Research Station of the Linze Farmland Ecosystem in Gansu Province. This station is located in the middle stream of the Heihe River and the southern edge of the Badain Jaran Desert. Its geographical coordinates are 39°21′N and 100°07′E (Figure 1). The terrain is flat, with an altitude of 1382 m. The main climatic characteristics are drought, high temperature, and windy, which belong to a typical temperate continental desert climate. The annual average precipitation is 119.16 mm, mostly concentrated in May–October, accounting for about 85% of the whole year. The relative humidity of the air is 46% and the annual evaporation is as high as 2390 mm, which is about 20 times that of precipitation [21] (Figure 2a,b). Wind and sand activities are strong and plant growth depends entirely on natural precipitation. The zonal soil is gray-brown desert soil, sandy loam soil, and sandy soil. The landform types are mainly fixed sandy land, semi-fixed sand dune, and mobile sand dune.

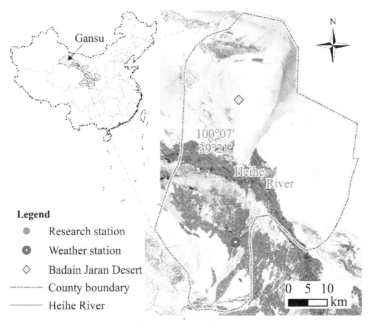

Figure 1. Location map of the study area.

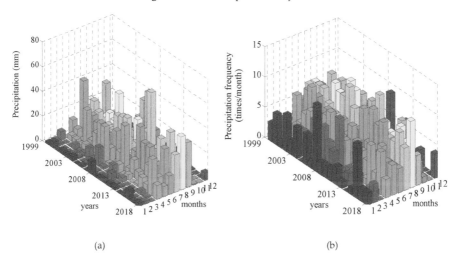

Figure 2. (**a**) Distribution of precipitation in Linze from 1999 to 2018; (**b**) precipitation frequency in Linze from 1999 to 2018.

2.2. Research Methods

2.2.1. Study Site and Experimental Design

The test site in the experimental field was relatively flat, well ventilated, and the soil was uniform in the same place with natural temperature. Before the experiment started, we measured the initial soil condition (nutrients, moisture et al.).The content of organic matter in the 0–30 cm soil layer is less than 0.5%, the total N and P content is relatively low and less than 0.05%, the total K content is less than 2.0%, the pH is about 9, and the water content of soil is relatively low, less than 4% on average. In early May 2017, we selected healthy seeds with uniform size and full grains for sowing, which were collected and stored from the previous year. *R. soongarica* seeds were soaked continuously in hot water at a temperature of 25 °C for 24 hours before being planted. Seeds were sown in rows at intervals of 0.3 m and a depth of 0.5–1.0 cm. After that, adequate irrigation was performed to ensure a certain emergence rate. In early June 2017, after the seedlings were well established, precipitation treatment began in 2018 while precipitation treatment began in late May.

The average annual precipitation in Linze has been 119.2 mm for many years. From 1999 to 2018, the overall precipitation remained stable, but the annual precipitation amount and precipitation frequency changed significantly (Figure 2a,b). The maximum and minimum annual precipitation is 186.3 mm (2010) and 71.1 mm (1999), which is 156.34% and 59.67% of the average annual rainfall of 119.2 mm. Except individual extreme years, in 90% of the past 20 years, annual precipitation is within the range of 70–130% of the average annual precipitation. The maximum and minimum monthly precipitation frequencies from July to October in the last 20 years were 9.3 times per month (2007) and 4 times per month (2004). From July to October, for many years, the average monthly precipitation frequency was 5.6 times per month. Therefore, in this experiment, three precipitation gradients, W (annual average precipitation), W− (W decreased by 30%), and W+ (W increased by 30%), were set with 119.16 mm as the control; two precipitation frequencies gradients were set, F (6 times per month) and F− (3 times per month); giving a total of 6 precipitation treatments: W−F, W− F− WF, WF−, W+F, and W+F−. The plot area was 3 × 2 m for each precipitation treatment, six plots for one block and three replicated blocks, with a total of one hundred and forty-four rows and about seven thousand and five hundred seeds (Figure 3).

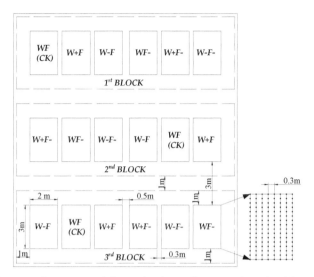

Figure 3. Diagram of experimental design for the seedling growth study. W, annual average precipitation; W+, 30% increase in precipitation; W−, 30% decrease in precipitation; F, annual average precipitation frequency (6 times per month); F−, 50% decrease in precipitation frequency (3 times per month). Black dots represent the random distribution of seedlings.

The temperature might be lower than 0°C before May, and sudden cooling might cause frostbite to the newly sprouted seedlings, thus affecting the experimental results. The research period was from May 2017 to October 2018. Except for the experimental precipitation treatment period, the seedlings were mainly irrigated by natural precipitation from the beginning of November 2017 to the end of April 2018. In the period of precipitation treatment, each block was provided with rain shelters, trenches, and ventilation around to keep other natural factors close to the natural conditions. The height of the canopy was 1.5 m and the sample plot extended around 1 m. It was fixed with wooden stakes. Shelters were removed on days without rain to minimize shelter effects on other environmental variables. During the whole experimental period, it was blocked at night, cloudy periods, and during precipitation to prevent natural precipitation from affecting the experiment. Artificial irrigation was performed according to the irrigation volume and irrigation frequency (Table 1) set in the experiment. Using a watering can, we evenly sprinkled the simulated precipitation amount set in the experiment in each plot. The irrigation time was from 19:00 to 20:00 on irrigation days (if the rain was absent for 1 day). From July to October, samples were taken on the 25th of each month, during which a total of 8 destructive sampling tests were performed. These selected data were for collecting samples from 2017 (25 July and 25 October) and 2018 (25 July and 25 October).

Table 1. Total number of precipitation events (treatments) in the *R. soongarica* growth experiments at two cycle levels from 2017 to 2018. W (annual average precipitation), W− (W decreased by 30%), and W+ (W increased by 30%).

Time	Average Monthly Precipitation (mm)	Event Size for Each Precipitation (mm)			Precipitation Cycle (d)	Precipitation Frequency (Times per Month)
		W−	W	W+		
July	26.7	3.1	4.5	5.8	5	6
		6.2	8.9	11.6	10	3
August	21.5	2.5	3.6	4.7	5	6
		5.0	7.2	9.3	10	3
September	18.0	2.1	3.0	3.9	5	6
		4.2	6.0	7.8	10	3
October	5.8	0.7	1.0	1.3	5	6
		1.4	1.9	2.5	10	3

2.2.2. Sampling Method

During sampling, we selected six healthy, uniform seedlings in each treatment. The seedling height was first measured with a steel tape measure; a vernier caliper was used to measure the base diameter, which was recorded. By using a spatula, the entire root system was carefully dug out and taken to the laboratory. The excess sand and soil were carefully removed; the length of the main root was measured with a tape measure. The above-ground and below−ground biomass were separated and dried in a 60 °C constant-temperature oven to obtain a constant weight. An electronic balance was used to weigh the above-ground biomass, below−ground biomass, and total biomass, the root/shoot ratio, the root/height ratio, the above-ground radial density, and the below−ground radial density.

2.2.3. Calculating the Root/Height Ratio, Above-Ground Radial Density, and Below−Ground Radial Density Parameters

Based on the fractal theory [22,23], we formulated the following morphological parameters and drew an abstract morphological figure of the seedlings (Figure 9):

$$RHR = \frac{H_R}{H_P}, \tag{1}$$

where RHR (main root/plant height ratio) is the root/height ratio; H_R is the main root length (cm); H_P is the plant height (cm);

$$ARD = \frac{M_A}{H_P}, \tag{2}$$

where ARD is the above-ground radial density ($g \cdot cm^{-1}$); M_A is the above-ground biomass (g); and H_P is the plant height (cm);

$$BRD = \frac{M_B}{H_R}, \tag{3}$$

where BRD is the below−ground radial density ($g \cdot cm^{-1}$); M_B is the below−ground biomass (g); and H_R is the main root length (cm).

2.2.4. Data Analysis

In different growth periods, two-factor analysis of variance distribution was used to evaluate the impacts of three precipitation treatments (W−, W, and W+), two precipitation frequency treatments (F and F−), and their interactions on plant height, basal diameter, main root length, above-ground biomass, below−ground biomass, total biomass, root/shoot ratio, root/height ratio, above-ground radial density, and below−ground radial density of *R. soongarica* seedlings. Least-significant difference

(LSD) was used to test for significant differences between samples. All analyses were performed using SPSS 21.0 and plots were performed using Matlab R2014a and Excel 2010.

3. Results

3.1. Impact of Precipitation Pattern Changes on Plant Height of R. Soongarica Seedlings

The impact of precipitation on plant height was significant ($P < 0.001$, Table 2). At both precipitation frequencies, plant height increased at W+ (Figure 4). When the precipitation increased by 30% (W+) and the frequency of precipitation was three times per month (F−), the plant height was the highest (Figure 4). Plant height increased with increasing precipitation and decreasing precipitation frequency. In 2017 (July and October) and 2018 (July and October), the same precipitation frequency (F) with increased precipitation (W+) increased plant height by 12.50%, 59.01%, 55.42%, and 71.93%, respectively. With the same amount of precipitation (W−, W, and W+), the difference in plant height was not significant at the two precipitation frequencies (F and F−) ($P > 0.05$, Figure 4).

Table 2. (F−values) based on two-way ANOVA of the impacts of precipitation amount and precipitation frequency on the indexes of *R. soongarica* seedlings.

Source of variation	Precipitation (W)	Precipitation frequency (F)	Precipitation (W)×frequency (F)
Plant height	29.669 (0.0000) ***	2.646 (0.1298)	0.179 (0.8386)
Basal diameter	1.286 (0.3119)	0.278 (0.6075)	0.019 (0.9812)
Main root length	3.730 (0.0550)	0.069 (0.7974)	0.041 (0.9600)
Above-ground biomass	14.267 (0.0007) ***	2.746 (0.1234)	0.010 (0.9897)
below−ground biomass	10.358 (0.0024) **	0.012 (0.9154)	0.100 (0.9054)
Total biomass	13.157 (0.0009) ***	0.861 (0.3717)	0.010 (0.9896)
Root/shoot ratio	1.371 (0.2909)	14.135 (0.0027) **	1.588 (0.2444)
Main root length/Plant height ratio (RHR)	134.23 (0.0000) ***	2.562 (0.1354)	0.876 (0.4416)
Above-ground radial density (ARD)	10.516 (0.0023) **	0.925 (.03551)	0.250 (0.7825)
below−ground radial density (BRD)	92.207 (0.0000) ***	1.069 (0.3216)	0.055 (0.9470)

Note: The different levels of probability considered are ***$P < 0.001$, ** $P < 0.01$, * $P < 0.05$. W, precipitation; F, precipitation frequency. Replicate number = 3.

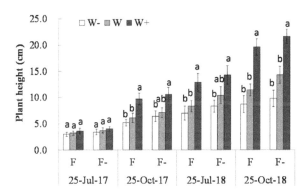

Figure 4. Dynamics of plant height of *R. soongarica* seedlings with different precipitation patterns from 2017 to 2018. For all plots, different lowercase letters are significantly different ($P < 0.05$) based on single-factor analysis of variance (ANOVA) with the same precipitation frequency. Bars indicate standard errors, n = 3.

3.2. Impact of Precipitation Pattern Changes on Above-Ground and Below–Ground Biomass and Total Biomass Accumulation

The impact of precipitation on the accumulation of above-ground, below–ground, and total biomass was significantly different ($P < 0.01$, Table 2). At both precipitation frequencies, above-ground and below–ground biomass and total biomass increased at W+ (Figure 5a–c). When the precipitation increased by 30% (W+) and the frequency of precipitation was three times a month (F−), the above-ground biomass and total biomass were the highest (Figure 5a,c). At the same precipitation frequency (F), an increase in precipitation (W+) promoted the accumulation of above-ground biomass, which increased by 94.59%, 101.38%, 54.93%, and 71.95%, respectively, during the growing season. The impact of increased precipitation (W+) on below–ground biomass was only significant in the growth of seedlings in July 2017 and October 2018 ($P < 0.01$, Figure 5), which were 108.04% and 68.53%, respectively, and the other growth periods were not significant ($P > 0.05$, Figure 5). The same amount of precipitation (W) with a reduction in precipitation frequency (F−) had a significant difference in seedling above-ground biomass with an increase of 33.06% in July 2018. There were no significant differences in other growth stages ($P > 0.05$). Changes in precipitation amount and precipitation frequency had significant impacts on the total biomass of *R. soongarica* seedlings; the trends were the same as those for the above-ground biomass. There was no significant difference in the impact of precipitation reduction (W−) on the below–ground biomass at the two frequencies (F and F−) ($P > 0.05$).

3.3. Impact of Precipitation Pattern Changes on The Root/Shoot Ratio

There was a significant difference in the impact of precipitation increasing on root/shoot ratio ($P < 0.05$) in October 2017 and July 2018, which decreased by 34.77% and 36.89%, respectively. There were not significant ($P > 0.05$, Table 2 and Figure 6) in other periods. While, there was not significantly different in the impact of precipitation decreasing on root/shoot ratio ($P > 0.05$) (except for July 2017). W+F treatmen decreased by 34.77% in October 2017 and W+F treatmen decreased by 36.89% in July 2018, respectively. The effect of precipitation frequency on root/shoot ratio was significantly different ($P < 0.01$, Table 2). Decrease in precipitation frequency (F−), W−F− treatment decreased by 60.15% in July 2017. W−F− treatment decreased by 24.60%, WF− treatment decreased by 26.47%, W+F− treatment decreased by 28.55% in October 2017, W+F− treatment increased by 28.57% in July 2018, and WF− treatment decreased by 31.75% in October 2018, respectively.

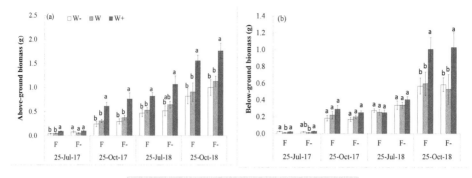

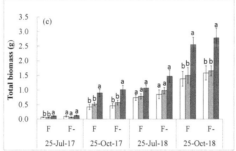

Figure 5. Dynamics of above-ground biomass (**a**), below–ground biomass (**b**) and total biomass (**c**) of *R. soongarica* seedlings with different precipitation patterns from 2017 to 2018. Different lowercase letters are significantly different ($P < 0.05$) based on single-factor analysis of variance (ANOVA) with the same precipitation frequency. Bars indicate standard errors, n = 3.

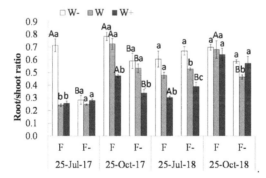

Figure 6. Dynamics of the root/shoot ratio of *R. soongarica* seedlings with different precipitation patterns from 2017 to 2018. Different capital letters indicate that in the same precipitation, the difference between the change of precipitation frequency and the control is significant ($P < 0.05$), and different small letters indicate that the difference between the change of precipitation and the control is significant ($P < 0.05$).

3.4. Impact of Precipitation Pattern Changes on the Root/Height Ratio

The impact of precipitation change on the root-to-height ratio is extremely significant ($P < 0.001$, Table 2 and Figure 7). From October 2017, the root-to-height ratio stabilized under various precipitation treatments. The main root/height ratio (RHR) was 5.435–6.639 under W– treatment, 2.750–3.694 under W treatment, and 1.676–2.439 under W+ treatment. When precipitation frequency (F) was constant, with an increase in precipitation (W+), seedling growth in July and October 2017 was not significantly different (*P* > 0.05), but it was significantly different in July and October 2018 (*P* < 0.05), decreasing by 37.96% and 50.01%, respectively. When the precipitation decreased (W−), the root/height ratio showed significant differences in each period (*P* < 0.001), increasing by 67.44%, 79.70%, 42.78%, and 74.80%, respectively. When the amount of precipitation was constant (W) with decreasing precipitation frequency (F−), the growth stage of the seedlings was not significantly different (*P* > 0.05).

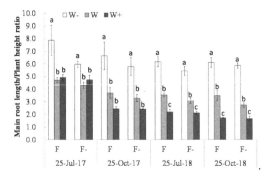

Figure 7. Dynamics of the main root length/plant height ratio of *R. soongarica* seedlings with different precipitation patterns from 2017 to 2018. Different lowercase letters are significantly different (*P* < 0.05) based on single-factor analysis of variance (ANOVA) with the same precipitation frequency. Bars indicate standard errors, n = 3.

3.5. Impact of Changes in Precipitation Patterns on Above-Ground Radial Density and Below−Ground Radial Density

The impact of precipitation changes on the above-ground and below−ground radial density was significantly different (*P* < 0.01, Table 2). The overall above-ground radial density showed an increase when precipitation increased in July and October 2017, but the difference was not significant in July 2018. The density increased when precipitation decreased in October 2018. When precipitation frequency (F) was constant, an increase of precipitation (W+) caused a significant difference in above-ground radial density in July and October 2017 (*P* < 0.001, Figure 8a), increasing by 72.96% and 26.64%, respectively. The difference between July and October was not significant (*P* > 0.05, Figure 8a). When precipitation decreased (W−), the above-ground radial density was significantly different in October 2018 (*P* < 0.001, Figure 8a), showing an increase of 74.80%.

When precipitation frequency (F) was constant, an increase in precipitation (W+) had a significant effect on the below−ground radial density of seedlings between July 2017 and October 2018 (*P* < 0.01), which increased by 82.51% and 93.12%, respectively. When precipitation decreased (W−), the differences in the below−ground radial density of seedlings between July and October 2017 and July 2018 were significant (*P* < 0.01, Figure 8b), which increased by 67.36%, and decreased by 45.65% and 24.28%, respectively. When precipitation was constant (W), a decrease in precipitation frequency (F−) had a significant effect on below−ground radial density in July 2018 (*P* < 0.01, Figure 8b), with an increase of 23.18%, and no significant difference at other times (*P* > 0.05, Figure 8b).

3.6. Abstract Morphological Figure of The Impact of Different Precipitation Patterns on the Growth of R. Soongarica Seedlings

When the precipitation increased by 30%, the plant height, aboveground biomass, and below−ground radial density (BRD) increased, and the root system was relatively thicker and shorter. However, when the precipitation decreased by 30%, the plant height, above-ground biomass, and below−ground radial density (BRD) decreased, and the whole root system was relatively thinner and longer. The maximum above-ground biomass was obtained in W+F− treatment (Figure 9).

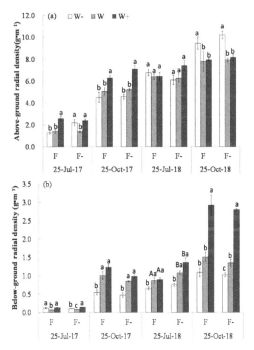

Figure 8. Dynamics of above-ground radial density (ARD) (**a**) and below−ground radial density (BRD) (**b**) of *R. soongarica* seedlings with different precipitation patterns from 2017 to 2018. Different lowercase letters are significantly different ($P < 0.05$) based on single-factor analysis of variance (ANOVA) with the same precipitation frequency. Bars indicate standard errors, n = 3.

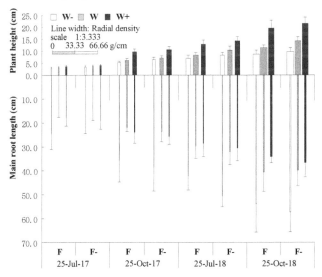

Figure 9. Abstract morphological figure of the impacts of different precipitation patterns on the growth of *R. soongarica* seedlings from 2017 to 2018. The radial density is expressed by the width of the line. The above-ground biomass is expressed by the rectangular area of the line width and height. The scale on the figure is only meaningful for the line width (radial density) and has no constraint on the plant height and the main root length.

4. Discussion

Climate change has changed the existing precipitation pattern and had a profound impact on ecosystems, especially in arid and semi-arid ecosystems where precipitation resources are scarce. The distribution of plant biomass in various organs can reflect the plant's adaptability to the environment and its growth strategy. The response of plants to changes in precipitation amount and frequency affects the future of desert ecosystems and regional sustainable development.

4.1. Impact of Precipitation on the Growth of R. Soongarica Seedlings

In desert ecosystems, water is a key factor for plant growth and survival [24]. Plant growth and development are the results of a combination of genetic factors and environmental conditions [25]. It is generally accepted that increased precipitation has a significant positive impact on plant growth in research on responses to increasing precipitation [26]. This study found that the increase of 30% in precipitation significantly increased the plant height, above-ground biomass, and total biomass of *R. soongarica*. This result is similar to that of Duan et al. and Shan et al. [18,20]. However, some studies have suggested that increasing precipitation does not significantly promote the growth of desert plants [10]. The results of this study are different from those of Sun [10], indicating that the *R. soongarica* seedlings are more sensitive to ecosystem disturbance than adults, especially in plant height. Rapid height growth in this species may have low net costs (and thus a high internal use efficiency) and would seem to favor minimizing allocation to lateral branches during juvenile stages. Li and Zhao [27] concluded that with an increase in precipitation, the plant height of *R. soongarica* seedlings decreased significantly, and the change in precipitation did not significantly affect their biomass and biomass allocation. This is different from the results of this study possibly because the precipitation and research objects of the two experimental settings are different. Li and Zhao [27] studied the seedlings that only grow for 20 days and the irrigation amount was 0.49–1.96 mm·d^{-1}, so that the water basically cannot effectively infiltrate. Our results are different because we studied 1-year and 2-year-old seedlings and had a higher irrigation frequency. There is a positive relationship between disturbance size or intensity and the availability of resources for plant growth. Whole plant growth should be strongly influenced by the factors such as the uptake of water and nutrients, the interception of light, and the allocation of carbon to the maintenance of roots and shoots.

This research found that the impact of precipitation changes on the *R. soongarica* below−ground biomass is different from the above-ground biomass. When precipitation is sufficient, water promotes the synthesis and transportation of auxin, which results in increased plant height, increased biomass, inhibited growth of the main roots below−ground, vigorous growth of lateral roots, and increased biomass. In the absence of precipitation, inhibition of auxin synthesis and transportation will slow plant growth, decrease above-ground biomass, cause vigorous growth of the main roots, and increase the below−ground biomass. Desert plants show different response strategies to different water conditions [28]. As a result of natural selection, physiological activities such as the distribution of plant roots in the arid zone have reached the level of maximum use of precipitation [29].

Studies have shown that both an increase and decrease in precipitation contribute to the below−ground biomass accumulation of desert plants [6]. This study found that the increase and decrease of precipitation in July 2017 (2 months of seedling growth) promoted the growth of the below−ground part (Figure 5). This result supports previous findings. However, the increase and decrease of water in October 2017 and July 2018 had no significant impact on the growth of the subsurface, which is basically the same as the result of Shan's study [20]. The increase in precipitation in October 2018 significantly promoted the accumulation of below−ground biomass and changed significantly from previous trends. This may be due to the fact that after rapid growth during the growing period, the plant has accumulated a large amount of biomass, while the water in October is relatively rare. Therefore, it is necessary to meet the water demand by inhibiting growth above ground (even becoming deciduous) and increasing the growth of the main roots in order to achieve a new water balance. Zhang's research found that the leaves of *R. soongarica* seedlings fell off early in the late stage of seedling growth (October), it was considered that the seedlings entered the dormant phase early, shortening the growth period [19]. This may be due to the lack of water, the adaptive mechanism of falling leaves in *R. soongarica* seedlings, and reducing evaporation, not necessarily entering the dormant period early. *R. soongarica* seedings adapt their resource supply mode by adjusting the changes of the above-ground parts mainly through the plant height and biomass.

4.2. Impact of Precipitation Frequency on the Growth of R. Soongarica Seedlings

In addition to total precipitation, water availability is also affected by soil characteristics, the temporal distribution of precipitation, such as the frequency, and the season of events [13]. Disturbance events frequency may be an important characteristic able to influence individuals' life cycles, the ecology and distribution of populations and of entire species, the structure and dynamics of community and of ecological processes [5]. Research has shown that in arid regions, reducing the frequency of precipitation and decreasing the amount of precipitation show a slight mutual inhibition [13]. However, the results of this study did not find that there was an interaction between precipitation amount and frequency of *R. soongarica* seedling growth (Table 2). During the whole growing season, the seedling plant height, above-ground biomass, and total biomass increased with the increase of precipitation. The treatment of decreasing precipitation frequency was slightly larger than the treatment of constant frequency, but the difference was not significant (Figures 4 and 5). The maximum plant height, above-ground biomass, and total biomass were obtained when the precipitation was increased by 30% and the frequency of precipitation was reduced by 50% (lower frequency, but individual precipitation is larger) (Figures 4 and 5). This shows that reducing the frequency of precipitation is better than the frequency-invariant treatment, but the impact is slight. The results are consistent with those of Schneider and Shan [13,20]. Gao et al. [11] believed that the values of all nutritional and reproductive traits in plants increase with increasing precipitation frequency. Our study found that the below−ground biomass of *R. soongarica* seedlings did not differ significantly at the two frequency levels. This is different from the results of Gao et al. [11], which may be because of different adaptation strategies. Gao et al.'s research object was one-year-old herbs and the net growth rates of herbs to an increase in water is different from that of shrubs [9].

4.3. Impact of the Interaction Between Precipitation Amount and Frequency on the Growth of R. Soongarica Seedlings

A series of complex factors such as precipitation, precipitation frequency, atmospheric evaporation, and soil structure characteristics control the interaction between precipitation amount and precipitation frequency to seeding, as well as their impacts on soil moisture content, and plant growth and recovery. These factors mean that there cannot be a universal law for large-scale precipitation time and plant water use [30]. Related research has also presented diversity. Some research [31] shows that the interaction between precipitation amount and precipitation frequency affects the main root length of (Nitraria spp), seedlings, above-ground biomass of *R. soongarica* seedlings [19], below−ground biomass, and total biomass [18]. There is an interaction between growth rates [20]. However, the results of this research did not find that there was an interaction between precipitation amount and frequency on *R. soongarica* seedling growth (Table 2), possibly because of different research objects. The research objects in this article were 1-year and 2-year-old *R. soongarica* seedlings, while the previous studies used 1-year-old *R. soongarica* seedlings, and even seedlings with a growth period of about 100 days. There are different growth strategies for different levels of water availability.

4.4. Abstract Morphological Figure of the Impacts of Different Precipitation Patterns on the Growth of R. Soongarica Seedlings

At present, scholars generally use plane graphics to accurately reflect the change range and trend of a single indicator, which is convenient for analysis and statistics. However, it is often difficult for a single indicator to reflect the overall and comprehensive traits of the research object, and it is impossible to comprehensively describe the correlation between plant morphological characteristics and plant growth. This paper introduces three morphological indicators: the main root length/plant height ratio (RHR), the above-ground radial density (ARD), and the below−ground radial density (BRD), combined with the plant height and main root length to draw an abstract figure of *R. soongarica* (Figure 9). The root/shoot ratio not only reflects the growth and coordination of the crops above and below the ground, but is also an important indicator of whether the crop can adapt to environmental factors (such as nutrients and water) [32]. However, the root/shoot ratio has limitations in reflecting plant morphology and evaluating environmental adaptability. This limitation is reflected in the fact that the root/shoot ratio can only reflect the distribution and coordination of the above-ground and below−ground crops from the biomass dimension, but cannot reflect the specific allocation patterns of above-ground and below−ground biomass. Ratios of dry mass fractions (e.g., root/shoot ratio; RMF) do not account for the more plastic response of tissue morphology, architecture, and physiology [33]. This is crucial because dry mass fractions can mask shifts in morphology or architecture by remaining constant [34]. The introduction of root/height ratio ((RHR) main root length to plant height ratio), above-ground radial density (ARD), and below−ground radial density (BRD) has expanded the research dimension, and it is possible to study how the plant biomass is distributed from the morphological dimension. The root system mainly absorbs water and nutrients from the soil for the growth and development of the entire plant. Current research generally believes that plants often increase the root/shoot ratio when water is scarce in order to use deep soil water to sustain growth [35]. With less water, the inhibition of root growth in shallow soils is enhanced, while the promotion in deep soils increases. The drier, the inhibition and promotion were stronger [36]. This study found that the root/shoot ratio decreased as precipitation increased overall, the root/shoot ratio did not increase significantly when water was reduced (except for July 2017) (Figure 6). This is inconsistent with general research results. However, this study found that the root-to-height ratio of seedlings responded significantly differently to water changes (Figure 7). The root/height ratio (RHR) increases with decreasing water content and decreases with increasing water content. This study found that the increase of precipitation in different periods had different effects on root-to-height ratio, the decrease of precipitation had no significant difference on root/shoot ratio ($P > 0.05$), and the decrease of precipitation frequency had significant difference on root/height ratio (RHR) ($P < 0.01$, Figure 7). This is inconsistent with the result of Duan et al. [18] where low rainfall promotes the increase of root/shoot ratio. It may be because there are differences in the research objects selected by Duan et al. [18]. The growth period was about 3 months, and the seedlings in the vigorous growth period were selected, while the research objects selected in this experiment were 1 year and 2 years from July 2017 to October 2018. The water condition and the performance of each growth stage were also different during this period. In addition, the root/shoot ratio was affected by multiple factors, and there was different performance in different growth periods. This study also found that the response of root/height ratio of seedlings to water changes was very significant (Figure 7). The root height ratio (RHR) increases with decreasing water content, decreases with increasing water content, and remains relatively stable. This is consistent with the results of studies by Kage, Xu, and Chen [37–39], where drought-tolerant plants increased root length/plant height ratio and reduced transpiration resistance to absorb water and adapt to a water deficit. The experimental data showed that the range of the root/height ratio of W−treatment is 5.435–6.639, W treatment is 2.750–3.694, and W + treatment is 1.676–2.439. This has showed that the root/height ratio of *R. soongarica* seedlings is more stable in characterizing the relationship between water and plant morphology.

Much research has been done on moisture in desert plants [7,20]. However, these research indicators (such as plant height, biomass, root/shoot ratio, root branch number, specific root length, etc.) are isolated from each other due to lack of connection, and cannot form a relatively comprehensive and systematic evaluation system. The introduction of above-ground radial density (ARD) and below−ground radial density (BRD) expands the traditional research dimension and helps to grasp and describe the plant morphology as a whole.

Above-ground radial density is the ratio of above-ground biomass to plant height and characterizes the general form of above-ground biomass distribution. When the water is sufficient, both the above-ground and below−ground plant can fully grow. The biomass is accumulated on the ground with high growth and increased leaves. The biomass accumulation rate is faster than the plant height growth rate and the above-ground radial density increases. The below−ground biomass accumulation rate is greater than the main root length growth rate and the below−ground radial density increases. When the water is insufficient, the growth of the above-ground plant is suppressed and the biomass is reduced. When the above-ground biomass accumulation rate is less than the plant height growth rate, the above-ground radial density decreases (the lower part mainly grows and absorbs water through the main root) and the below−ground biomass accumulates. When the growth rate is less than the main root, the below−ground radial density decreases and vice versa. This study found that firstly, above-ground radial density increased then remained constant as precipitation increased, while above-ground radial density decreased first and then increased with decreasing precipitation. This indicates that the above-ground bioaccumulation rate is greater than the plant height growth rate and less than the plant height growth rate as the precipitation increases, while the above-ground bioaccumulation rate is less than the main root length growth rate and greater than the main root length growth rate as the precipitation decreases.

Diversity in root morphology has declined sharply; the roots have become thinner across the sequence of tropical, temperate, and desert biomes, presumably owing to changes in resource supply caused by seasonally inhospitable abiotic conditions [40]. Comas et al. [34] found that in an environment with water deficits, small xylem diameters in seminal roots save soil water deep in the soil profile and capacity for deep root growth, and large xylem diameters in deep roots may also improve root acquisition of water when ample water at depth is available. The below−ground radial density is the ratio of the below−ground biomass to the length of the main root, and represents the general shape of the below−ground unit length biomass distribution. The results of this study found that when the precipitation increased, the below−ground radial density increased, and the root system was relatively thicker and shorter, while when the precipitation decreased, the below−ground radial density decreased, and the overall root system became finer and longer (Figures 8 and 9). This has enabled them to markedly improve their efficiency of soil exploration per unit of carbon invested and to reduce their dependence on symbiotic mycorrhizal fungi [40]. This is the same result as Comas and Ma [34,40]. Some results suggest that root traits have evolved along a spectrum bounded by two contrasting strategies of root life: an ancestral 'conservative' strategy in which plants with thick roots depend on symbiosis with mycorrhizal fungi for soil resources and a more-derived 'opportunistic' strategy in which thin roots enable plants to more efficiently leverage photosynthetic carbon for soil exploration [40]. The results of this study indicate that the *R. soongarica* seedlings' root adaptive strategy manifests as an obvious 'opportunistic' strategy.

5. Conclusions

Our findings suggest that in the context of climate change, the derivation of morphological traits has been important for preparing plants to colonize new habitats and for the rich generation of biodiversity within and across biomes; to the degree that a particular set of morphological and physiological traits will result in high growth efficiency for only a limited range of disturbance sizes or resource availability. The traits that determine *R. soongarica* growth appear to be more specialized in their response to disturbance. *R. soongarica* seedings adapt their resource supply mode by adjusting the changes of the above-ground parts mainly through the plant height and biomass and the length and thickness of the root system of the below−ground part. An improved functional understanding of morphological traits is critical for comprehending the distribution of plant life, and may help to predict the risk of species extinction and to conserve biodiversity, and to improve regional sustainable development capabilities in the face of environmental change.

Sustainability **2020**, *12*, 2439

Author Contributions: Conceptualization, Y.X.; Data curation, Y.X. and R.M.; Formal analysis, Y.X.; Investigation, T.X., R.M. and Z.Z.; Methodology, T.X.; Resources, Yi Li; Software, Z.Z.; Supervision, Y.L.; Writing – original draft, Y.X.; Writing – review & editing, Y.L. All authors have read and agreed to the published version of the manuscript.

Funding: This research was funded by The Innovation Base and Talent Plan of Gansu Provincial Science and Technology Department grant number (17JR7WA018), the Science and Technology Achievement Transformation Project of Gansu Provincial Education Department grant number (2017D-14) and the Scientific Research Project of Higher Education Institutions in Gansu Province grant number (2016A-026).

Acknowledgments: We thank the Linze National Field Station for Farmland Ecosystem, Chinese Academy of Sciences, for providing test site, the reviewers and editors for valuable suggestions that significantly improved this manuscript and MDPI for its linguistic assistance during the revision of this manuscript.

Conflicts of Interest: The authors declare no conflict of interest.

References

1. Qin, D.H. Climate change science and sustainable development. *Prog. Geogr.* **2014**, *33*, 874–883. (In Chinese)
2. Westra, S.; Fowler, H.J.; Evans, J.P.; Alexander, L.V.; Berg, P.; Johnson, F.; Kendon, E.J.; Lenderink, G.; Roberts, N.M. Future changes to the intensity and frequency of short-duration extreme rainfall. *Rev. Geophys.* **2014**, *52*, 522–555. [CrossRef]
3. Densmore-Mcculloch, J.A.; Thompson, D.L.; Fraser, L.H. Short-term effects of changing precipitation patterns on shrub-steppe grasslands: Seasonal watering is more important than frequency of watering events. *PLoS ONE* **2016**, *11*, e0168663. [CrossRef] [PubMed]
4. Burke, E.J.; Brown, S.J.; Christidis, N. Modeling the Recent Evolution of Global Drought and Projections for the Twenty-First Century with the Hadley Centre Climate Model. *J. Hydrometeorol.* **2006**, *7*, 1113–1125. [CrossRef]
5. Battisti, C.; Poeta, G.; Fanelli, G. *An Introduction to Disturbance Ecology*, 1st ed.; Springer Nature Switzerland AG: Zug, Switzerland, 2016; pp. 13–136.
6. Zhang, L.M.; Liu, X.P.; Zhao, X.Y.; Zhang, T.H.; Yue, X.F.; Yun, J.Y. Response of sandy vegetation characteristics to precipitation change in Horqin sandy land. *Acta Ecol. Sin.* **2014**, *34*, 2737–2745. (In Chinese)
7. He, J. Eco-Physiology Responses and Adaptive Strategies of Desert Species *Nitraria tangutorum* to Simulated Rain Addition. Ph.D. Dissertation, Chinese Academy of Forestry, Beijing, China, 2015. (In Chinese).
8. Sun, B.; Qian, J.P.; Zhao, H.X. Response of biomass and root morphology of desert plants *Corispermum candelabrum* to precipitation change in northwest China. *Ecol. Environ. Sci.* **2018**, *27*, 1993–1999. (In Chinese)
9. Gherardi, L.A.; Sala, O.E. Enhanced precipitation variability decreases grass- and increases shrub-productivity. *Proc. Natl. Acad. Sci. USA* **2015**, *112*, 12735–12740. [CrossRef]
10. Sun, Y.; He, M.Z.; Wang, L. Effects of precipitation control on plant diversity and biomass in a desert region. *Acta Ecol. Sin.* **2018**, *38*, 2425–2433. (In Chinese)
11. Gao, R.; Yang, X.; Liu, G.; Huang, Z.; Walck, J.L. Effects of rainfall pattern on the growth and fecundity of a dominant dune annual in a semi-arid ecosystem. *Plant Soil* **2014**, *389*, 335–347. [CrossRef]
12. Zeppel, M.J.B.; Wilks, J.V.; Lewis, J.D. Impacts of extreme precipitation and seasonal changes in precipitation on plants. *Biogeosciences* **2014**, *11*, 3083–3093. [CrossRef]
13. Schneider, A.C.; Lee, T.D.; Kreiser, M.A.; Nelson, G.T. Comparative and interactive effects of reduced precipitation frequency and volume on the growth and function of two perennial grassland species. *Int. J. Plant Sci.* **2014**, *175*, 702–712. [CrossRef]
14. Gibson-Forty, E.V.J.; Barnett, K.L.; Tissue, D.T.; Power, S.A. Reducing rainfall amount has a greater negative effect on the productivity of grassland plant species than reducing rainfall frequency. *Funct. Plant Biol.* **2016**, *43*, 380–391. [CrossRef]
15. Trenberth, K.E.; Dai, A.; Rasmussen, R.M. The changing character of precipitation. *Bull. Am. Meteorol. Soc.* **2003**, *84*, 1205–1218. [CrossRef]
16. The American Heritage Dictionary. Available online: https://www.thefreedictionary.com/Seedlings (accessed on 1 March 2020).
17. Fay, P.A.; Schultz, M.J. Germination, survival, and growth of grass and forb seedlings: Effects of soil moisture variability. *Acta Oecologica* **2009**, *35*, 679–684. [CrossRef]
18. Duan, G.F.; Shan, L.S.; Li, Y.; Zhang, Z.Z.; Zhang, R. Effects of changing precipitation patterns on seedling growth of *Reaumuria soongorica*. *Acta Ecol. Sin.* **2016**, *36*, 6457–6464. (In Chinese)

19. Zhang, Z.Z.; Shan, L.S.; Li, Y. Prolonged dry periods between rainfall events shorten the growth period of the resurrection plant *Reaumuria soongorica*. *Ecol. Evol.* **2018**, *8*, 920–927. [CrossRef]

20. Shan, L.S.; Zhao, W.Z.; Li, Y.; Zhang, Z.Z.; Xie, T.T. Precipitation amount and frequency affect seedling emergence and growth of *reaumuria soongarica* in northwestern China. *J. Arid Land* **2018**, *10*, 574–587. [CrossRef]

21. Liu, B.; Zhao, W.Z. Ecological Adaptability of Photosynthesis and Water Metabolism for *Tamarix Ramosissima* and *Nitraria Sphaerocarpa* in Desert-Oasis Ecotone. *J. Desert Res.* **2009**, *29*, 101–107. (In Chinese)

22. Mandelbrot, B.B. *The Fractal Geometry of Nature/Revised and Enlarged Edition*; WH Freeman and Company: New York, NY, USA, 1982; p. 460.

23. Lin, Q.D. Architectural Form-Finding Based on Fractal Theory. Ph.D. Dissertation, Tsinghua University, Beijing, China, 2014. (In Chinese).

24. Wu, Y. Physiological Response of Desert Plants in the Southern Margin of Junggar Basin to Light Rainfall Events. Ph.D. Dissertation, Chinese Academy of Sciences University, Urumqi, China, 2014. (In Chinese).

25. Yin, M.H.; Li, Y.N.; Zhou, C.M. Compensation effects of regulated deficit irrigation and tillering interference to winter wheat. *J. Appl. Ecol.* **2015**, *26*, 3011–3019. (In Chinese)

26. Xiao, C.W.; Zhou, G.S.; Zhao, J.Z. Effect of different water conditions on growth and morphology of *Artemisia ordosica* Krasch. seedlings in Maowusu sandland. *Acta Ecol. Sin.* **2001**, *21*, 2136–2140. (In Chinese)

27. Li, Q.L.; Zhao, W.Z. Responses of seedlings of five desert species to simulated precipitation change. *J. Glaciol. Geocryol.* **2006**, *28*, 414–420. (In Chinese)

28. Zhou, H.; Zhao, W.Z.; He, Z.B. Water sources of *Nitraria sibirica* and response to precipitation in two desert habitats. *Chin. J. Appl. Ecol.* **2017**, *28*, 2083–2092. (In Chinese)

29. Fay, P.A.; Carlisle, J.D.; Knapp, A.K.; Blair, J.M.; Collins, S.L. Productivity responses to altered rainfall patterns in a C_4-dominated grassland. *Oecologia* **2003**, *137*, 245–251. [CrossRef] [PubMed]

30. Deng, W.P. Water Use Mechanism of Typical Tree Species in Beijing Mountainous Areas. Ph.D. Dissertation, Beijing Forestry University, Beijing, China, 2015. (In Chinese).

31. Zhang, R.; Shan, L.S.; Li, Y. Effect of change to simulated precipitation patterns on seedling growth of *Nitraria tangutorum*. *Acta Prataculturae Sin.* **2016**, *25*, 117–125. (In Chinese)

32. Xu, G.W.; Wang, H.Z.; Zhai, Z.H.; Sun, M.; Li, Y.J. Effect of water and nitrogen coupling on root morphology and physiology, yield and nutritionutilization for rice. *Trans. Chin. Soc. Agric. Eng.* **2015**, *31*, 132–141. (In Chinese)

33. Boot, R.G.A.; Mensink, M. Size and morphology of root systems of perennial grasses from contrasting habitats as affected by nitrogen supply. *Plant Soil* **1990**, *129*, 291–299. [CrossRef]

34. Comas, L.H.; Becker, S.R.; Cruz, V.M.V.; Byrne, P.F.; Dierig, D.A. Root traits contributing to plant productivity under drought. *Front. Plant Sci.* **2013**, *4*, 442. [CrossRef]

35. Guo, J.P.; Gao, S.H. Response of sandland plants growth to CO_2 enrichment and soil drought. *J. Soil Water Conserv.* **2004**, *18*, 174–176. (In Chinese)

36. Liang, Q. Response of Growth and Physiological Characteristics to Soil Depth Under Different Water Conditions on Lolium Perenne in Karst Region. Master's Dissertation, Southwest University, Chongqing, China, 2016. (In Chinese).

37. Kage, H.; Kochler, M.; Stützel, H. Root growth and dry matter partitioning of cauliflower under drought stress conditions: Measurement and simulation. *Eur. J. Agron.* **2004**, *20*, 379–394. [CrossRef]

38. Xu, B.; Shan, L. A Study comparing water use efficiency and root/ shoot ratio of alfalfa and *Astragalus adsurgens* at seedling stage. *Acta Agrestia Sin.* **2003**, *11*, 78–82. (In Chinese)

39. Chen, S.Y. Effect of Water Stress on the Growth and Quality of Alfalfa and Its Physiological Basis. Master's Dissertation, Shandong Agricultural University, Taian, China, 2006. (In Chinese).

40. Ma, Z.; Guo, D.; Xu, X.; Lu, M.; Bardgett, R.D. Evolutionary history resolves global organization of root functional traits. *Nature* **2018**, *555*, 94–97. [CrossRef] [PubMed]

Article

A Quantitative Analysis of Socio-Economic Determinants Influencing Crop Drought Vulnerability in Sub-Saharan Africa

Bahareh Kamali [1,*], Karim C. Abbaspour [1], Bernhard Wehrli [2] and Hong Yang [1,3]

[1] Eawag, Swiss Federal Institute of Aquatic Science and Technology, 8600 Duebendorf, Switzerland; karim.abbaspour@eawag.ch (K.C.A.); hong.yang@eawag.ch (H.Y.)
[2] Institute of Biogeochemistry and Pollutant Dynamics, ETH Zurich, 8092 Zurich, Switzerland; bernhard.wehrli@eawag.ch
[3] Department of Environmental Sciences, University of Basel, 4056 Basel, Switzerland
* Correspondence: bahareh.kamali@eawag.ch

Received: 4 September 2019; Accepted: 31 October 2019; Published: 3 November 2019

Abstract: Drought events have significant impacts on agricultural production in Sub-Saharan Africa (SSA), as agricultural production in most of the countries relies on precipitation. Socio-economic factors have a tremendous influence on whether a farmer or a nation can adapt to these climate stressors. This study aims to examine the extent to which these factors affect maize vulnerability to drought in SSA. To differentiate sensitive regions from resilient ones, we defined a crop drought vulnerability index (*CDVI*) calculated by comparing recorded yield with expected yield simulated by the Environmental Policy Integrated Climate (EPIC) model during 1990–2012. We then assessed the relationship between *CDVI* and potential socio-economic variables using regression techniques and identified the influencing variables. The results show that the level of fertilizer use is a highly influential factor on vulnerability. Additionally, countries with higher food production index and better infrastructure are more resilient to drought. The role of the government effectiveness variable was less apparent across the SSA countries due to being generally stationary. Improving adaptations to drought through investing in infrastructure, improving fertilizer distribution, and fostering economic development would contribute to drought resilience.

Keywords: Crop model; EPIC; regression techniques; drought adaptation

1. Introduction

The current world population has already reached 7 billion, and by 2050 it is estimated to increase to 9 billion, with the most significant increase concentrated in South Asia and Sub-Saharan Africa (SSA) [1,2]. A number of studies show that if no drastic actions are taken in consumption patterns, shifting diets, and reducing wastes, meeting the world's food demand will require doubling food production and this will unavoidably exert extreme pressure on agricultural and natural resources [3]. On the other hand, climate change has exacerbated the situation by increasing drought risks and flood events around the globe with significant impacts on agricultural production. As agriculture in most SSA countries is primarily rainfed, SSA is at the core of this threat. Achieving food security in SSA is an enormous challenge given the currently weak institutions, poor infrastructure, population, unfavorable weather conditions, and high social dependence on declining natural resources [4]. Therefore, to better cope with the consequences of droughts and mitigate their effects, it is essential to incorporate socio-economic information in the analysis of drought vulnerability in SSA.

Drought is a natural disaster, and the degree of its impacts is well recognized in terms of the magnitude of vulnerability. Various definitions have been proposed for vulnerability.

The Intergovernmental Panel on Climate Change (IPCC) defined vulnerability as a function of exposure, sensitivity, and adaptive capacity [5]. The fifth assessment report focused on socio-economic aspects and prioritizing adaptation interventions. Adaptive capacity is the ability of a system to cope with the consequence of drought and represents the potential to implement measures that help reduce potential impacts [6]. Quantification of adaptation capacity is a challenge for two reasons. First, adaptation is the intrinsic property of a system and only manifests itself when the system is exposed to a shock [7]. Second, adaptation is heterogeneous over space and time, and therefore, it is challenging to generalize a set of universal factors that enhance adaptive capacity in different regions over time.

Several studies have attempted to incorporate adaptive capacity in the vulnerability assessment using approaches such as aggregated quantitative indicators [8–10] or semi-structured interviews [11–13]. Indicators of adaptive capacity were also grounded in sustainable livelihood theory as a set of different forms of assets to which people have access, such as financial, human, resource, or physical assets [14]. This approach calculates a composite index for each constituent asset. During the last two decades, many studies also attempted to employ the sustainable livelihood concept in their analysis of vulnerability [13,15,16].

The use of indicators is one of the most common ways to define factors influencing vulnerability. Despite the extensive research, improvement is required to identify and choose driving forces and indicators that more accurately determine the relationship between socio-economic factors and vulnerability [17]. Detailed studies along these lines have been recently conducted at the European [7,11,18], Australian [13,16,19], and in few cases at global scales [8,20], but the issue has not been thoroughly addressed in SSA countries. Naumann et al. [21] calculated composite indicators that reflect different aspects of vulnerability and adaptive capacity at the Pan-African level. Their study, however, lacks an empirical and analytical framework that can clearly explain the direction and magnitude of effectiveness (or influence) of individual factors on vulnerability. This limitation impedes policy development to mitigate the vulnerability. Several studies have used regression models to analyze the relationship between the socio-economic variables and crop drought vulnerability in some individual countries or regions in SSA [22,23]. However, these studies are case-specific, and it is challenging to upscale their findings to the whole of SSA.

In assessing the factors influencing vulnerability to drought, most studies have not considered the temporal variations in different countries. Multivariable regression models used for these analyses often take the prevailing situation of individual countries over a specified time period. In reality, vulnerability is time-dependent and varies with the severity of the drought. It is, therefore, important to consider changes in vulnerability and factors that influence it on temporal dimension. As data is increasingly becoming available in developing countries, panel data analysis at the national level provides a more significant basis for modeling the complexity of socio-economic factors influencing vulnerability [24]. Panel analysis is a statistical method to analyze two-dimensional data collected over time for the same entity. However, such an analysis exceeding a single cross-section of time has so far not been conducted in the context of crop drought-vulnerability assessment in SSA.

In the literature, most studies quantifying drought vulnerability are based on some arithmetic aggregation of several representative components. Very few studies have quantified the drought vulnerability by comparing the expected crop yield under certain climate conditions with the recorded yield. The magnitude of the difference between modeled and recorded yield is an indicator of resiliency of a region to drought occurrence. Process-based crop models are practical tools to model how yield changes in response to climate variability or drought. Linking the drought vulnerability defined as the distance between the expected and recorded yield with socio-economic variables may, therefore, enhance the choice of management options.

The aim of this study is to identify those socio-economic variables that contribute to drought-vulnerability of maize production in SSA. To this end we: (1) construct a definition of Crop Drought Vulnerability Index (*CDVI*) obtained by comparing simulated yield from a crop model with recorded yield at the country level; (2) compare the adaptive capacity variables for economic, human, resource, infrastructure, and governance categories across countries; (3) apply multivariable panel regression to identify socio-economic variables influencing *CDVI* for maize; and finally (4) discuss the implications of the results for mitigation of maize vulnerability to drought in SSA.

2. Materials and Methods

2.1. Site Description

SSA is home to over 1 billion people. The average annual precipitation of 795 mm yr^{-1} is diversely distributed in different regions [25]. Small landholders depend on rainfed agriculture as their primary source of livelihood. The effects of population increase and climate change have exacerbated the risk of hunger [26]. Therefore, reducing crop vulnerability to drought is essential for this region. Maize is the most widely grown crop and staple food in SSA [27] and represents a relevant case for this study. In the last two decades, average maize yields in SSA have increased from around 1.4 to 1.8 t ha^{-1}, with over 40% increase in South Africa, but they are still at the very bottom of globally reported maize yields [28].

2.2. Quantifying Drought Exposure Index

Drought exposure index, used in this study to separate drought from non-drought years, was obtained from the Standardized Precipitation Index (*SPI*) [29]. To obtain *SPI*, first, a suitable cumulative probability distribution function (here Gamma distribution) [30,31] is fitted to the precipitation. We chose a two-parameter gamma distribution as the cumulative distribution function [30,31]. *SPI* is then obtained from applying the inverse normal function with a mean of 0 and a standard deviation of 1 to the cumulative distribution function. After this transformation, 99.7% of data vary between −3 and 3. Negative values are representative of drought situations, whereas the positive values show non-drought cases (Table 1). The five classifications within the [−3, 3] range are defined as wet, near normal, mild drought, moderate drought, and severe drought [30] (Table 1).

Table 1. Five categories of the Standardized Precipitation Index (*SPI*) and Crop Drought Vulnerability index (*CDVI*). *SPI* is selected to separate drought and non-drought years [30].

SPI Classes	*CDVI* Classes	*SPI* or *CDVI*
Wet	No drought vulnerability	1.0 and more
Near normal	Low vulnerability	0.001 to 0.99
Mild	Mild vulnerability	−0.99 to 0.0
Moderate	High vulnerability	−1.499 to −1.0
Severe	Very high vulnerability	−1.50 or less

The analysis was carried out for years 1990–2012. Precipitation was obtained from the WFDEI (WATCH-Forcing-Data-ERA-Interim) meteorological forcing data [32] at 0.5° resolution. The grid level precipitation data were aggregated to the national level using weighted areal average of maize cultivated lands obtained from MIRCA2000 version 1.1 [33]. *SPIs* were calculated at different time scales. We tested the suitability of 1, 3, 6, 9, and 12-month time scales. For each country, we selected the time scale with the highest correlation coefficient between *SPI* and simulated maize yield during the growing season.

2.3. Definition of Crop Drought Vulnerability Index (CDVI)

We focused on maize vulnerability to drought occurrences. The definition of maize *CDVI* follows the study of Bryan, et al., [13], which is the difference between the recorded and modeled yields. In this study, the impact of drought on yield was quantified using the EPIC (Environmental Policy Integrated Climate) crop model. EPIC simulates different processes of farming systems as well as their interactions using data such as weather, soil, landuse, and crop management parameters [34].

EPIC operates on a daily time step and is capable of simulating crop growth under various climatic and environmental conditions and complex management schemes [34].

EPIC is a field-scale model. Following the work done by Kamali et al., [35], its application was extended to larger scales using a Python framework named EPIC$^+$. The framework divides the region of study into several grids based on a specified resolution (here 0.5°) and executes EPIC on each grid cell. EPIC$^+$ is also equipped with the Sequential Uncertainty Fitting (SUFI-2) [36] algorithm for calibration. More details on the procedure used to simulate and calibrate maize using EPIC$^+$ are found in the work of Kamali, et al., [35].

The yield simulated by EPIC$^+$ reflects the influence of climate variability on crop production and, therefore, represents the expected yield (Y^{Exp}). The actual yield (Y^{Act}) is obtained from FAO, which is reported at the national level [28]. The recorded yield is influenced by both climate factors and many socio-economic factors. We defined the variable Y as $Y_t{}^{Act} - Y_t{}^{Exp}$. If Y^{Act} is larger than Y^{Exp} (positive Y), then the region is considered to be resilient to changing climate conditions, and so the vulnerability is low. However, if Y^{Exp} is larger than Y^{Act} (negative Y), then the maize production is sensitive to climate variability, and consequently the vulnerability is high. Finally, if Y^{Act} is equal to Y^{Exp} (Zero Y), then the region has effective adaptation. For better spatial and temporal comparisons of Y across SSA, we normalized it using *Z-score* transformation calculated as [37]:

$$CDVI_t = \frac{Y_t - Mean(Y)}{STD(Y)} \tag{1}$$

where Y_t is the yield residuals for year t i.e., $Y_t = Y_t{}^{Act} - Y_t{}^{Exp}$; $Mean(Y)$ and $STD(Y)$ are the mean and standard deviation of the yield residuals (Y), respectively. Similarly, and using this transformation, $CDVI$ can also be classified into five levels of vulnerability severity (Table 1).

2.4. Selecting Socio-Economic Variables Relating to CDVI

To select the variables influencing *CDVI*, we used the following step-wise approach:

Step (1): Candidate socio-economic variables were collected as recommended by Brooks, et al., [38] and Naumann, et al., [21]. The two studies together introduced over 50 socio-economic variables that might be important for climate vulnerability assessments at the national scale. We selected 17 drought-relevant variables with more than 50% data availability for the study period (1990–2012) (Table 2). For the ease of referencing, we classified the variables into five categories as economic, human, resource, infrastructure, and governance. To fill the missing values of the variables, we used the spline interpolation procedure applied by Simelton, et al., [9]. For variables such as fertilizer use, where temporal data was missing for a certain period, we used the average of the years with available data.

Step (2): The variables were expressed in a variety of ranges or scales. In this step, they were transformed into a uniform dimension. We use the *Z-score* transformation as the normalization technique calculated as the ratio of the residual of the variable and its mean divided by the standard deviation as [39].

$$Nvariable_{t,i} = \frac{variable_{t,i} - Mean(variable_i)}{STD(variable_i)} \tag{2}$$

where $Mean(variable_i)$ and $STD(variable_i)$ are the mean and standard deviation of variable i ($i = 1, 2, \ldots, 17$) in year t. Equation (2) converts all variables to a common scale with an average of 0 and standard deviation of 1. It is also consistent with the standardized scale of *CDVI*. For those variables such as "Interest payment" (Table 2), which are negatively correlated to *CDVI*, the inverse of values were used (i.e., 1/variable) in normalization. We did not transform variables of the governance category as these variables were obtained from a combination of different criteria that were already normalized during the procedure of quantifying these variables.

Table 2. Potential socio-economic variables influencing drought vulnerability and their definitions (variables are defined per year).

Category	Variable	Definition	Unit
Economic	GDP/capita	Gross domestic product	US$/capita
	Interest payment	Interest payments on external debt	% GNI
	GNI	Gross national income	US$/capita
	Agriculture GDP	Agriculture GDP	% GDP in total GDP
Human	HDI	Human development index	ratio ranging from 0 to 1
	Health expenditure	Health expenditure per capita	US$/capita
	Maternal mortality	Maternal mortality ratio	per 100,000 live births
	Calorie intake	Calorie intake per capita	calorie
	Food production index	Food crops that are edible and contain nutrients excluding coffee and tea. (average of 2004–2006 equals 100)	ratio of each year to the base period (2004–2006)
Resource	Agricultural area (ha/capita)	per capita land area that is either arable, under permanent crops, or under permanent pastures	ha/capita
	Fertilizer use (t/ha)	Nitrogen fertilizer use	tons/ha
Infrastructure	Water access (%)	% of population with access to improved drinking water source	percentage
	Electricity access (%)	% of rural population with access to electricity	percentage
Governance	Control of corruption	The extent to which public power is exercised for private gain, including both petty and grand forms of corruption, as well as "capture" of the state by elites and private interests	normalized values ranging from −2.5 to 2.5
	Government effectiveness	The quality of public and civil service, policy formulation and implementation, the degree of its independence to political pressures, the credibility of the government's commitment to policies	normalized values ranging from −2.5 to 2.5
	Political stability	The likelihood that the government will be destabilized or overthrown by unconstitutional or violent means	normalized values ranging from −2.5 to 2.5
	Voice & accountability	The extent to which a country's citizens are able to participate in selecting their government, freedom of expression, freedom of association, and a free media	normalized values ranging from −2.5 to 2.5

Sources: The variables in the first four categories were obtained from the World Bank (http://www.worldbank.org/). The variables for the "governance" category were obtained from Kaufmann et al., [40] and can be downloaded from www.govindicators.org.

Step (3): The influence of 'multicollinearity' is an essential issue in regression models which can seriously distort the interpretation of a model [41]. This is related to situations when predictors are correlated with each other. Therefore, some factors will be redundant. To avoid the redundancy of variables, a bivariate correlation matrix was constructed between *Nvariable*s of each category. This helped us to evaluate the strength and direction of the linear relationships between the variables [39]. The statistically most significant variables were selected, and the rest were removed from further assessment.

Step (4): The selected variables of each country were averaged to obtain an aggregated value for the economic, human, resource, infrastructure, and governance categories. This facilitates the spatial comparison of the selected aspects and provided insights into the adaptive capacity of different countries.

Step(5): Statistical regression models were constructed to analyze the role of selected variables in Step 3 for characterizing *CDVI* (see next Section 2.5 for details).

2.5. Linking CDVI and Socio-Economic Variables Using Regression Models

Two model formulations were constructed, and their results were compared to get some insights into the influence of the temporal dimension on explanatory variables. In both models, the socio-economic variables selected in Step 3 were considered as explanatory variables and *CDVI* as a dependent variable. The first model (Model I) did not take into account temporal variability of explanatory and dependent variables. Only their averages were included. The aim of this model was to check if the temporally aggregated variables (during drought years) can explain the relationship between socio-economic conditions and maize vulnerability. The model I is expressed as:

$$\text{Model I}: \overline{CDVI}_i = \alpha_0 + \beta \overline{x}_i + \varepsilon_i \qquad (3)$$

where $\overline{CDVI}_i$ and $\overline{x}$ stand for average *CDVI* and selected socio-economic variables for country i during drought years respectively, α_0 is a constant intercept term, β is a vector of coefficient for each explanatory variable, and ε_i is the error term. The model is fitted with all socio-economic variables and simplified by comparing versions with and without a particular explanatory term using the Likelihood Ratio tests. In each step, we remove insignificant terms until all remaining factors are significant at 1%, 5%, or 10% levels

Model II takes into account the temporal dimension of socio-economic factors using the panel regression method. A Linear Fixed-effect model is designed to analyze the impact of variables that vary over time. The formulation for Model II is expressed as:

$$\text{Model II}: CDVI_{z,t} = \alpha_i + \beta x_{i,t} + \varepsilon_{i,t} \qquad (4)$$

where $CDVI_{z,t}$ is the dependent variable at time t for country z, and x stands for selected socio-economic variables as the dependent variables. In this model, each country has its characteristics that may or may not influence the predictor. The α_i values take into account the individual effects of country z that are not explained by the dependent variables included in Model II. Explanatory variables were added one by one in a forward stepwise manner to check their significance. The model was then selected by comparing versions with and without a particular explanatory term using Likelihood Ratio tests. In each step, we selected a model in which all factors are significant at 1%, 5%, or 10% levels. MATLAB ver2018 was used to implement the statistical regression models.

3. Results

3.1. Temporal and Spatial Patterns of CDVI

The *SPI*-based spatial and temporal distribution of yearly drought exposure during 1990 and 2012 showed many drought events in all SSA countries with higher severity from 1990 to 1999 as compared to the period 2000–2012 (Figure 1). Between 1990 and 1995, almost all countries experienced drought events. The most severe intensity in Southern Africa occurred in 1992, in Central Africa during 1994–1995, and in Western Africa in 1990. From 1996 to 2012, less severe drought events were recorded. The severe droughts were only observed in countries of Western Africa in 2002 and Eastern Africa in 2004. Other countries mostly experienced mild to moderate droughts (Figure 1).

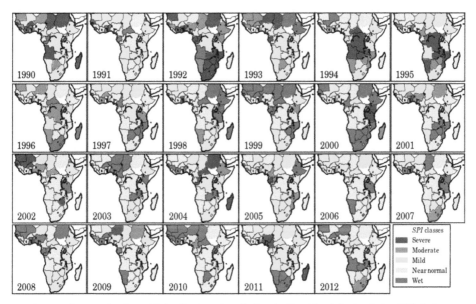

Figure 1. Country-level spatial distribution of yearly Standardized Precipitation Index (*SPI*) during 1990–2012; the classification is based on Table 1 definition.

CDVI calculated based on the difference between actual and expected yield identified sensitive and resilient countries during 1990 and 2012 (Figure 2). The expected yield was obtained from the crop model, only reflecting climate-induced stresses. By contrast, the recorded yield showed the influence of actual climate as well as socio-economic conditions. Countries with high vulnerability had negative *CDVI*s, whereas resilient ones exhibited positive values. The yearly national level *CDVI*s showed many severe to moderate intensities of vulnerability during 1990–2012.

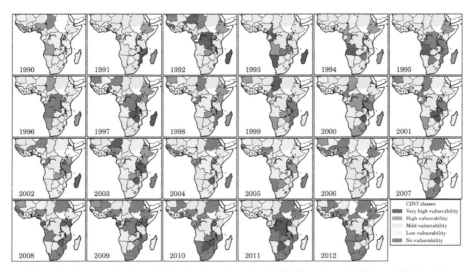

Figure 2. Country level spatial distribution of yearly *CDVI* (Crop Drought Vulnerability Index) calculated based on the difference between the actual and expected yields; the classification is based on Table 1 definitions.

During 1990–2012, *CDVI* showed similar temporal trends as *SPIs*, as both indices showed more severe drought and higher vulnerability during 1990–1999 as compared to 2000–2012. However, the two indices displayed different pictures in terms of annual severity, revealing that different countries had varying resiliency during drought events. The Southern African countries were less affected by extreme *CDVI*, especially after 2000, indicating higher resiliency. For example, the moderate drought in 2012 in South Africa (Figure 1) did not cause the same level of vulnerability (Figure 2). The 2011 drought, which occurred in most SSA countries (Figure 1), had different vulnerabilities in different countries (Figure 2). An opposite situation occurred in 1999 when a moderate vulnerability in Western Africa was seen despite a non-drought situation.

Overall, based on *SPI* classification, Southern African countries, as well as Kenya, Tanzania, and Ethiopia from Eastern Africa, Mali, Niger, Nigeria, Botswana, Chad, and Central African Republic were more prone to severe droughts whereas Central African countries and Madagascar experienced less severe droughts due to the occurrence of higher rainfall (Figure 3a). However, *CDVI* showed a different picture as South Africa, Botswana, Mozambique, Nigeria, and Cameroon were less vulnerable (Figure 3b). The severity of vulnerability was lower than the severity of drought in countries such as Kenya, Tanzania, and Mali.

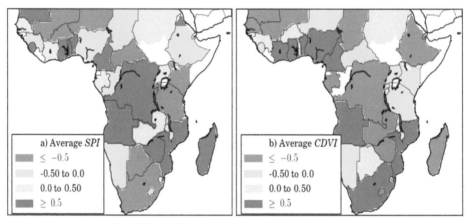

Figure 3. Country-level comparison of (**a**) average *SPI* and (**b**) average *CDVI* during 1990 to 2012.

3.2. Socio-Economic Factors Influencing CDVI

The bivariate correlation coefficient, i.e., the '*r*' values between normalized variables were calculated to identify variables that are significantly correlated to each other (Step 3). Overall, nine socio-economic variables were selected (Table 3). In the economic category, the highest correlation coefficient was found between "GDP/capita" and "Interest payment" ($r = 0.64$), between "GDP/capita" and "GNI" ($r = 0.83$), and between "Interest payment" and "GNI" ($r = 0.59$) (Figure S1 in supplementary materials). This shows that these three variables have similar effects. Therefore, we selected "GDP/capita" and "Agricultural GDP (%GDP)" as the final variables from the economic category.

The human category showed the highest correlation coefficient (Figure S2 in supplementary materials). For example, we found $r = 0.84$ between "HDI" and "Health expenditure" and $r = 0.73$ between "Calorie intake" and "Food production index" (Figure S2 in supplementary materials). "HDI" was selected as the most representative variable, as it is a composite statistic of life expectancy, education, and per capita income and encompasses a more general aspect of human development. "Food production index" was selected due to the significant correlation coefficient with other explanatory variables from the human category. It was also an indicator of human nutrition status and a representative of human health aspects (definition in Table 2). While "HDI" and "Food production index" were also correlated, we selected both at this step and decided to check the suitability of one or both for the regression model. We retained both variables in the resource and infrastructure categories (Water resource access (%) and Electricity access (%)) due to their relevance for drought vulnerability. In the governance category, the correlation coefficient between variables was larger than 0.75 (Figure S3 in supplementary materials). We only kept "Government effectiveness" as it is a more general indicator and encompasses political, rules, and regulatory aspects [40] (Table 2).

The nine remaining variables varied significantly from one region to the other in terms of their absolute values (Table 3). For example, variables such as "GDP/capita", "HDI", "Fertilizer use (t/ha)" were significantly larger in Southern SSA at the 5th, 50th, and 95th percentiles. "Agricultural GDP (%GDP)" was significantly larger in Central Africa, while variables such as "Food production index" showed approximately similar percentiles in the four SSA regions (Table 3).

Table 3. The nine socio-economic variables in the five categories (economic, human, resource, infrastructure, and governance) selected after the pairwise correlation coefficient analysis (Step 3) (units of each variable are shown in Table 1).

Category	Variable	5th, 50th, 95th Percentiles			
		Eastern SSA	Southern SSA	Central SSA	Western SSA
Economic	GDP/capita	177, 330, 700	696, 3025, 4455	224, 911, 6060	253, 551, 924
	Agriculture GDP	16, 36, 50	3.4, 8, 11.6	7.2, 23.2, 52.3	6.7, 34, 48.5
Human	HDI	0.3, 0.39, 0.48	0.46, 0.6, 0.9	0.3, 0.4, 0.6	0.26, 0.39, 0.49
	Food production index	82, 93, 101	89, 95, 103	87, 90, 108	82, 88, 97
resource	Agricultural area	0.23, 0.99, 3.4	0.53, 3.2, 5.8	1.1, 2.2, 21	0.43, 0.88, 12.3
	Fertilizer use	1.7, 5.7, 34	3.1, 7.8, 58	0.35, 3.5, 8.9	0.62, 7.8, 31
Infrastructure	Water resource access	32, 55, 78	54, 79, 95	45, 63, 84.6	44, 63, 83
	Electricity access	0.9, 2.6, 16	2.0, 17, 47	1, 13, 31	1, 8.2, 25
Governance	Government effectiveness	−1.4, −0.7, −0.4	−0.7, 0.2, 0.6	−1.8, −1.2, −0.6	−1.3, −0.8, −0.1

The temporal variability of the nine selected and normalized variables showed different patterns for different variables and countries. For "GDP/capita", almost all SSA countries, except Zimbabwe, Gambia, and Guinea, showed an increasing trend with a rather steep slope after 2000 (Figure 4). "Agricultural GDP (%GDP)" exhibited different temporal variability across different countries. Both "HDI" and "Food production index" showed an increasing trend all over SSA except Zimbabwe for "HDI" and Democratic Republic of Congo (DRC) for "Food production index". The data for temporal variability of "fertilizer use (t/ha)" was not available for the period 1990–2000 and we used the average of the years 2001–2012. From the two infrastructure variables, "Water resource access (%)" showed similar trends and values in all countries except Zimbabwe and Sudan, but "Electricity access (%)" was more variable across countries. For "Government effectiveness", the large variability between countries masked any temporal trends (Figure 4).

To see the spatial variability of selected socio-economic variables across countries in each category, we aggregated variables of each category into one indicator by calculating the average of their normalized values (Step 4). In all five categories, Southern African countries had higher values (higher adaptive capacity). In the economic category (Figure 5a), Western and Eastern SSA showed approximately similar lower values. Other countries like Zimbabwe, Zambia, and Angola were placed in between with aggregated values between −0.35 and 0. The human aspect showed low values in Ethiopia, Angola, Niger, and Mali followed by most Western African countries especially Angola and Namibia (Figure 5b). In the resource category, mostly Central Africa showed the lowest values (between −0.85 and −0.35) (Figure 5c). The infrastructure showed a different picture with mostly low values for most Central and Eastern African countries (Figure 5d). All SSA countries showed very poor capacity in terms of Governance, with the exception of Botswana and South Africa (Figure 5e).

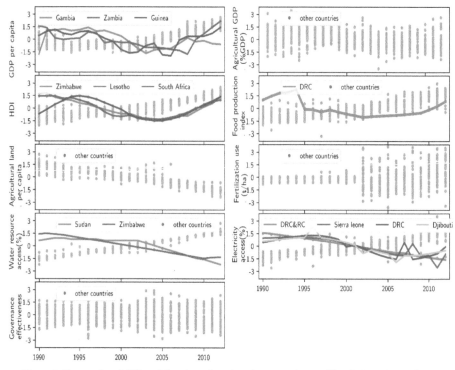

Figure 4. Temporal variability of the nine selected socio-economic variables during 1990 and 2012; the variables are normalized using *Z-score* in SSA (DRC: Democratic Republic of Congo and RC: Republic of Congo). Each scatter point is representative of a specific country. The black line highlights countries with different temporal variability.

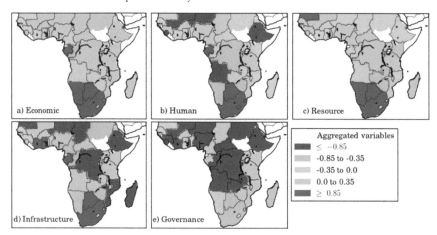

Figure 5. The spatial variability of aggregated indicators in five categories: (**a**) economic, (**b**) human, (**c**) resource, (**d**) infrastructure, and (**e**) governance obtained from the average of selected normalized socio-economic variables in each category.

3.3. Relations between Time-Invariant Socio-Economic Variables and CDVI

We calculated the average of nine selected socio-economic variables (as explanatory variables) and *CDVI* (as dependent variables) during drought years, i.e., *SPI* < 0 (Step 5). All variables were fixed at this step, and a multiple variable regression model was used. The model was constructed (Model I) to check the suitability of the mean independent and dependent variables for model explanation and then to determine which of the nine selected socio-economic variables can explain the vulnerability of maize to drought. The relatively low R^2 value of 0.30 (Table 4) indicates that the time-invariant model is probably not sufficiently explanatory. Of the nine variables, only three including "Agricultural GDP (%GDP)", "Food production index", and "Electricity access (%)" were identified as statistically significant variables for reducing the vulnerability to drought. The highest β value for "Agricultural GDP (%GDP)" with a value of −0.72 indicated that this variable was the most influential for reducing vulnerability followed by "Electricity access (%)" (β = 0.34) and "Food production index" (β = 0.32).

Table 4. The time-invariant socio-economic factors influencing maize drought vulnerability obtained from Model I. The averages of socio-economic variables and *CDVI* were used in the analysis. Only variables that were significant at 1%, 5%, or 10% levels were included in the model. The empty rows (-) pertain to those that were not significant (*SE*: standard error) (units of each variable are shown in Table 1).

Model I: Simple linear regression model using the mean of 11 selected variables: R^2 = 0.30				
Variable	β	SE	t-stat	P-values
Intercept	−1.25	0.79	−1.58	0.9
GDP/capita	-	-	-	-
Agriculture GDP	−0.72	0.25	−2.87	**0.007**
HDI	-	-	-	-
Food production index	0.32	0.16	1.99	**0.054**
Agricultural land	-	-	-	-
Fertilizer use	-	-	-	-
Water resource access	-	-	-	-
Electricity access	0.34	0.18	1.90	**0.065**
Government effectiveness	-	-	-	-

3.4. Assessing the Relationship between Time-Variant Socio-Economic Variables and CDVI

The temporal variability of socio-economic variables (as explanatory variables) and *CDVI* (as dependent variables) was examined during 1980–2012. As variables had both time and space (individual country) dimensions, panel regression models were used. We tested and compared two different models. In the first model (Model IIa), variables during drought and non-drought years (1980–2012) were included, while in the second model (Model IIb), only drought years were selected. The main reason for comparing these two models was to check if influential factors differ depending on climate stresses. Besides, drought has a long term effect, and the influence of some factors might be revealed in non-drought years. We believe that this comparison provides more detailed insights into the influencing factors of drought vulnerability.

We started with variables in the economic category i.e., "GDP/capita" and "Agricultural GDP (%GDP)" and then added variables of other categories one by one in a forward stepwise manner to check their significance. The comparison was made using the Likelihood Ration test. The results show significant improvement as the R^2 values were 0.63 in Model IIa and 0.60 in Model IIb (Table 5) over that of Model I (R^2 = 0.3). Model IIa slightly outperformed Model IIb. The *SE* values of the socio-economic variables were slightly lower compared to Model I in Table 4. This indicates that Models IIa,b are better able to identify factors that are important to maize vulnerability to drought. Including temporal variability of indicators is, therefore, critical to better characterize the effects of socio-economic factors on vulnerability mitigation.

Table 5. The time-variant socio-economic factors influencing maize drought vulnerability obtained from the fixed-effect panel data regression model (Model II). The time series of socio-economic variables were used in the analysis. Only variables that were significant at 1%, 5%, or 10% levels were included in the model. The empty rows (-) pertain to those that were not significant (*SE*: standard error) (units of each variable are shown in Table 1).

Model IIa: Fixed-effect regression model based on including drought and non-drought years: $R^2 = 0.66$				
Variable	β	SE	t-stat	P-values
GDP/capita	-	-	-	-
Agriculture GDP	−0.09	0.032	−2.87	<0.001
HDI	0.073	0.049	1.47	0.011
Food production index	1.94	0.20	9.66	<0.001
Agricultural land	-	-	-	-
Fertilizer use	0.132	0.055	2.401	0.02
Water resource access	-	-	-	-
Electricity access	2.787	0.642	4.34	<0.001
Government effectiveness	0.22	0.148	1.64	−0.01

Model IIb: Fixed-effect regression model based on only including drought years: $R^2 = 0.68$				
	β	SE	t-stat	P-values
GDP/capita	-	-	-	-
Agriculture GDP	−0.11	0.047	−2.30	0.02
HDI	0.158	0.070	2.39	0.019
Food production index	2.01	0.299	6.72	<0.001
Agricultural land	−0.109	0.047	−2.306	0.020
Fertilizer use	0.141	0.083	2.24	0.019
Water resource access	-	-	-	-
Electricity access	3.101	1.08	2.87	<0.001
Government effectiveness	-	-	-	-

A comparison of influential variables of Model I with Models IIa,b shows that more factors were significant for vulnerability in Models IIa,b. Apart from the three significant variables of Model I ("Agriculture GDP (%GDP)", "Food production index", and "Electricity access (%)"), "HDI", "Agricultural land (ha/capita)", and "Fertilizer use (t/ha)" were significant in Models IIa,b at 1%, 5%, or 10% levels. "Government effectiveness" was only significant in Model IIa, and "Agricultural land (ha/capita)" was only significant in Model IIb. Besides, the influential factors of Model I are significant at the 1% level (higher confidence), indicating their importance for vulnerability mitigation. All other variables were significant at the 5% level. In both Models, IIa,b, "Food production index" and "Electricity access (%)" showed higher β values (respectively, 1.94 and 2.787 in Model IIa; 2.01 and 3.101 in Model IIb).

Comparison of Model IIa with Model IIb showed that "Government effectiveness" was only statistically significant in Model IIa when all years were included. The β value for "Fertilizer use (t/ha)" was slightly higher in Model IIb (with a value of 0.141) compared with Model IIa (with a value of 0.132), indicating that as drought is occurring, the influence of fertilizer application is becoming more critical as an adaptation strategy. Besides, the β values of the human category, i.e., "HDI" and "Food production index" were higher, suggesting that their influence might be more critical during drought (Table 5). The two variables of "GDP/capita", and "Water resource access (%)" were not significant in the three models. The country-level intercept (α) values obtained from two fixed-effect models (Model IIa and Model IIb) show different values for different countries and also in two models (Table S1).

4. Discussion

4.1. Changes in the Crop Drought Vulnerability

In this study, we assessed the relationship between maize drought vulnerability and socio-economic variables influencing vulnerability at the national level for SSA countries during 1990–2012. We identified significant socio-economic factors that predispose an area's maize harvest to be resilient or sensitive to drought. The residuals of the simulated and observed yields were identified as prominent indicators of resilience concerning drought vulnerability. The response of a crop model adequately reflects the dynamic of climate on the rainfed maize yield. This is advantageous over studies that use third or fourth-order regression modeling to simulate yield [9,37], as such a de-trending procedure does not reflect the undeniable influence of climate variability on yield.

The overall spatiotemporal patterns of drought and vulnerability resemble the insights that different countries are coping differently with the occurring drought. An interesting observation is that maize vulnerability to drought has become less severe in recent years of our study period. For example, while the drought exposure is higher in Eastern and Southern African countries due to lower amounts of rainfall after 2000, the vulnerability of maize yield has declined. This corroborates the results of Naumann et al., [21], who concluded that these regions were less vulnerable. It suggests that there might be generic socio-economic factors that help mitigating vulnerability in different regions. We should emphasize that the period of our analysis excluded extreme drought events that occurred in the late 1980s over the entire SSA. Including this period might significantly improve the assessment of changes in *CDVI* and factors influencing the trend. However, since most socio-economic variables are not available before the 1990s, this period was not included, which is a limitation of this study.

This paper presents the first comprehensive vulnerability assessment at a large scale by linking a physically-based spatially explicit crop model with indicators of adaptive capacity. The study is unique in the current literature, as most analyses on agriculture vulnerability have been case-specific and done at national or community levels [42–45]. Other large-scale analyzes lack the integration of a biophysical crop models with socio-economic factors [46,47]. Despite the challenges, complexity, and associated uncertainty, we made the best use of the existing data by taking into account both temporal and spatial dimensions of socio-economic components to have a broader understanding at the continental level.

4.2. Major Factors Influencing Drought Vulnerability

The 17 potential socio-economic variables in the five categories helped to disclose different aspects of adaptive capacities and the related potential for reducing vulnerability. While this classification is not ideal and some indicators may fall in more than one category, it gives more details on adaptive capacity of different factors. We also mapped the spatial and temporal dynamics of selected variables. Out of 17 variables, 9 remained after the 'multicollinearity' analysis from which 3–7 variables were significant in the regression models depending on drought and non-drought years included in the study. The following conclusions can be drawn for the variables in each of the five categories:

Economic category: "GDP/capita" was not a statistically significant factor. This is probably because the definition of HDI already encompasses the economic status of a country. "Agricultural GDP (%GDP)", as a more specific variable for agriculture, was associated with high vulnerability of maize in all models (Model I, Model IIa, and Model IIb) at 1% or 5% levels. The negative sign of β coefficient indicates that the large share of "Agricultural GDP (%GDP)" in total GDP can have a significant effect on reducing the ability to cope with drought. A strong economy secures the system by facilitating implementation of coping strategies against environmental risk and drought exposure. It also provides possibility for higher investment in weather forecasting, which may help farmers to better prepare for drought. A weak economy, often represented by the large share of agricultural sector in GDP, has the opposite effect [9,17,48].

Human category: "Food production index" was identified as a statistically significant factor in both time-variant and time-invariant models with a relatively high positive β coefficient. The index is a measure of food security, which is consistent with the situation in SSA where maize is one of the staple crops, and its vulnerability can significantly influence human health status in Africa. It is also representative of economic conditions of a country. The effect of "HDI" in reducing vulnerability becomes apparent when time-variant factors are included. This suggests that more investment in increasing life expectancy and education (as components of "HDI") will reduce vulnerability. Bahadur Kc et al., [12] also showed that increasing "HDI" can result in an additional 6.8 million tons of maize production at the global level and the increase is more remarkable for developing countries such as Africa.

Resource category: Three variables were representative of the system's natural resources from which only "Agricultural land (ha/capita)" and "Fertilizer use (t/ha)" had significant roles. Both variables were more influential during drought periods (higher β coefficients in Model IIa and Model IIb). This suggests that fertilizer is increasingly used to mitigate the impact of drought on maize. The lower β coefficient of "fertilizer use (t/ha)" compared to other variables might be related to missing values for the period 1990–2000 when most significant droughts occurred. We used the average of 2001–2012 in the model to substitute the missing data. Obtaining more accurate data for fertilizer application in SSA will help to better understand the potential benefits. The statistical significance of "Agricultural land (ha/capita)" indicates that the population growth together with the limited land resources will be a significant threat to maize-based food security in the future.

Infrastructure category: "Water resource access (%)" showed no statistical significance in all models. This might be due to the temporal variation of this variable (in Figure 4), which shows only a linear increasing trend with no significant variability across countries and inter-annually. By contrast, "Electricity access (%)" with more regional variability (Figure 4) and with high β coefficients in all models is a better representative.

Governance category: "Governance effectiveness" was also identified as a critical factor for reducing vulnerability in Model IIa. As mentioned by Keshavarz et al., [49], adaptation at a governmental level can help create a set of practical long-term plans and policies which may enhance the capacity to develop, revise, and execute drought policies. The "Governance effectiveness" was less significant in Model IIb because the missing values for the period of 1990–1995 (when extreme droughts occurred) were filled with constant mean values. Therefore, no significant variability was noticed in the variable "Governance effectiveness" during extreme drought years. Hence, a deeper implication of this variable requires data at higher spatial and temporal resolutions.

4.3. Comparison of the Models Explaining the Relationship between CDVI and Socio-Economic Variables

We tested the suitability of time-variant and time-invariant variables and concluded as expected that including the temporal dimension of variables was necessary for the determination of socio-economic factors influencing drought. Drought is a time-dependent phenomenon and is characterized by climate conditions for a specified period. Panel regression was the method of choice to evaluate the influence of socio-economic variables. As socio-economic variables differ significantly from one year to another, aggregating the severity of multiple drought years by calculating their average disguises the influence that a specific socio-economic variable might have in a specified year. Implementing panel data results in a more accurate inference of model parameters as these models usually have more degrees of freedom and more sample variability.

5. Conclusions and Limitation

Overall, our results underline the suitability of regression models for identifying how socio-economic factors influence drought-affected maize production between 1990–2012 in SSA. Despite the usefulness, some limitations to the data used in this study call for caution in the interpretation and further empirical efforts to improve data quality. Our spatial scale lacks details at the sub-national level as socio-economic data and maize yield were reported at the national level. As also mentioned by Conway et al., [50] and Simelton et al., [51], regional or gridded data could identify which regions contribute most to national food insecurity. The World Bank databases reporting country level data were the only available sources with time series of socio-economic variables. Another limitation of this study is related to the lack of high-quality data on socio-economic variables. Official statistics in many SSA countries are not reliable and subject to criticism [52–54]. Therefore, there is still urgent need to invest in improving data quality.

There are also some other factors such as pest or disease that might have resulted in harvest loss and therefore increased crop drought vulnerability. However, we did not have access to these types of crop failure data. Such levels of information will depend on farm-level surveys. Another limitation is that the selected explanatory variables are not crop-specific or even agriculture-specific. Other factors, such as heat or cold spells, might have influenced crop vulnerability. However such specifications demand more work at the farm scale, which takes into account other drivers of vulnerability.

In conclusion, the current study was a preliminary but novel effort in identifying influential socio-economic factors on drought vulnerability across SSA. The results and the approaches developed can be used as a baseline study for further research to analyze crop drought vulnerability and its mitigation. As the quality and resolution of the data improve, a better understanding of the interaction of variables and their effects on drought vulnerability will be achieved by upgrading the calibrated crop model and also updating the analyses through inclusion of more recent years.

Supplementary Materials: The following are available online at http://www.mdpi.com/2071-1050/11/21/6135/s1, Figure S1: The visual representation of scatter plots generating correlation coefficient between normalized variabls of the economic category; Figure S2: The visual representation of scatter plots generating correlation coefficient between normalized variables of the human category; Figure S3: The visual representation of scatter plots generating correlation coefficient between normalized proxies of the governance category; Table S1: The coutnry level intercept (α) values obtained from two fixed effect models (Model IIa and Model IIb).

Author Contributions: Conceptualization, H.Y. and B.K. methodology, B.K.; software, K.C.A.; validation, B.W., K.C.A. and B.K.; formal analysis, B.K.; investigation, B.K.; resources, H.Y.; data curation, B.K.; writing—original draft preparation, B.K.; writing—review and editing, B.K., B.W., H.Y. and K.C.A.; visualization, B.K.; supervision, B.W., H.Y.and K.C.A.; project administration, Swiss Federal Institute for Aquatic Science and Technology (Eawag); funding acquisition, H.Y.

Funding: Swiss National Science Foundation.

Acknowledgments: We are thankful to the Swiss National Science Foundation (SNF No.: CR21I3_146430. Dec.2013-Nov.2016) for financial support.

Conflicts of Interest: All co-authors of this manuscript declare that they have no conflict of interest.

References

1. Lipper, L.; Thornton, P.; Campbell, B.M.; Baedeker, T.; Braimoh, A.; Bwalya, M.; Caron, P.; Cattaneo, A.; Garrity, D.; Henry, K.; et al. Climate-smart agriculture for food security. *Nat. Clim. Chang.* **2014**, *4*, 1068–1072. [CrossRef]

2. United Nations. *World Population Prospects: The 2015 Revision, Key Findings and Advance Tables*; Department of Economic and Social Affairs, Population Division: United Nations, 2015.

3. Foley, J.A.; Ramankutty, N.; Brauman, K.A.; Cassidy, E.S.; Gerber, J.S.; Johnston, M.; Mueller, N.D.; O'Connell, C.; Ray, D.K.; West, P.C.; et al. Solutions for a cultivated planet. *Nature* **2011**, *478*, 337–342. [CrossRef] [PubMed]

4. Webber, H.; Gaiser, T.; Ewert, F. What role can crop models play in supporting climate change adaptation decisions to enhance food security in Sub-Saharan Africa? *Agric. Syst.* **2014**, *127*, 161–177. [CrossRef]

5. IPCC. *Climate Change 2014: Impacts, Adaptation, and Vulnerability*; Cambridge University Press: United Kingdom and New York, NY, USA, 2014.

6. Adger, W.N. Vulnerability. *Glob. Environ. Chang.-Hum. Policy Dimens.* **2006**, *16*, 268–281. [CrossRef]

7. Williges, K.; Mechler, R.; Bowyer, P.; Balkovic, J. Towards an assessment of adaptive capacity of the European agricultural sector to droughts. *Clim. Serv.* **2017**, *7*, 47–63. [CrossRef]

8. Yeni, F.; Alpas, H. Vulnerability of global food production to extreme climatic events. *Food Res. Int.* **2017**, *96*, 27–39. [CrossRef] [PubMed]

9. Simelton, E.; Fraser, E.D.G.; Termansen, M.; Benton, T.G.; Gosling, S.N.; South, A.; Arnell, N.W.; Challinor, A.J.; Dougill, A.J.; Forster, P.M. The socioeconomics of food crop production and climate change vulnerability: A global scale quantitative analysis of how grain crops are sensitive to drought. *Food Secur.* **2012**, *4*, 163–179. [CrossRef]

10. Fraser, E.D.G.; Simelton, E.; Termansen, M.; Gosling, S.N.; South, A. "Vulnerability hotspots": Integrating socio-economic and hydrological models to identify where cereal production may decline in the future due to climate change induced drought. *Agric. For. Meteorol.* **2013**, *170*, 195–205. [CrossRef]

11. Blauhut, V.; Stahl, K.; Stagge, J.H.; Tallaksen, L.M.; De Stefano, L.; Vogt, J. Estimating drought risk across Europe from reported drought impacts, drought indices, and vulnerability factors. *Hydrol. Earth Syst. Sci.* **2016**, *20*, 2779–2800. [CrossRef]

12. Bahadur Kc, K.; Legwegoh, A.F.; Therien, A.; Fraser, E.D.G.; Antwi-Agyei, P. Food price, food security and dietary diversity: A comparative study of urban Cameroon and Ghana. *J. Int. Dev.* **2017**, *30*, 42–60.

13. Bryan, B.A.; Huai, J.; Connor, J.; Gao, L.; King, D.; Kandulu, J.; Zhao, G. What actually confers adaptive capacity? Insights from agro-climatic vulnerability of Australian wheat. *PLoS ONE* **2015**, *10*, e0117600. [CrossRef] [PubMed]

14. Ellis, F. The determinants of rural livelihood diversification in developing countries. *J. Agric. Econ.* **2000**, *51*, 289–302. [CrossRef]

15. Keshavarz, M.; Maleksaeidi, H.; Karami, E. Livelihood vulnerability to drought: A case of rural Iran. *Int. J. Disaster Risk Reduct.* **2017**, *21*, 223–230. [CrossRef]

16. Huai, J.J. Integration and typologies of vulnerability to climate change: A case study from Australian wheat sheep zones. *Sci. Rep.* **2016**, *6*, 33744. [CrossRef] [PubMed]

17. Vincent, K. Uncertainty in adaptive capacity and the importance of scale. *Glob. Environ. Chang.-Hum. Policy Dimens.* **2007**, *17*, 12–24. [CrossRef]

18. Blauhut, V.; Gudmundsson, L.; Stahl, K. Towards pan-European drought risk maps: Quantifying the link between drought indices and reported drought impacts. *Environ. Res. Lett.* **2015**, *10*. [CrossRef]

19. Huai, J.J. Dynamics of resilience of wheat to drought in Australia from 1991–2010. *Sci. Rep.* **2017**, *7*. [CrossRef]

20. Ericksen, P.; Thornton, P.; Notenbaert, A.; Cramer, L.; Jones, P.; Herrero, M. *Mapping Hotspots of Climate Change and Food Insecurity in the Global Tropics*; Research Program on Climate Change Agriculture and Food Security (CCAFS) Report No. 5; Climate Change Agriculture and Food Security (CCAFS): Copenhagen, Denmark, 2011.

21. Naumann, G.; Barbosa, P.; Garrote, L.; Iglesias, A.; Vogt, J. Exploring drought vulnerability in Africa: An indicator based analysis to be used in early warning systems. *Hydrol. Earth Sys. Sci.* **2014**, *18*, 1591–1604. [CrossRef]

22. Epule, T.E.; Ford, J.D.; Lwasa, S. Projections of maize yield vulnerability to droughts and adaptation options in Uganda. *Land Use Policy* **2017**, *65*, 154–163. [CrossRef]

23. Gbetibouo, G.A.; Ringler, C.; Hassan, R. Vulnerability of the South African farming sector to climate change and variability: An indicator approach. *Nat. Resour. Forum* **2010**, *34*, 175–187. [CrossRef]

24. Hsiao, C. Panel data analysis—Advantages and challenges. *Test* **2007**, *16*, 1–22. [CrossRef]

25. Ward, C.S.; Torquebiau, R.; Xie, H. *Improved Agricultural Water Management for Africa's dryland*; the World Bank: Washington, DC, USA, 2016.

26. Iglesias, A.; Quiroga, S.; Diz, A. Looking into the future of agriculture in a changing climate. *Eur. Rev. Agric. Econ.* **2011**, *38*, 427–447. [CrossRef]

27. Folberth, C.; Yang, H.; Gaiser, T.; Liu, J.G.; Wang, X.Y.; Williams, J.; Schulin, R. Effects of ecological and conventional agricultural intensification practices on maize yields in Sub-Saharan Africa under potential climate change. *Environ. Res. Lett.* **2014**, *9*, 044004. [CrossRef]

28. FAO. FAOSTAT Crop Statistical Database, Food and Agricultural Organization of the UN, Rome. 2012. Available online: http://www.fao.org/faostat/en/#data/QC (accessed on 1 June 2017).

29. McKee, T.B.; Doesken, N.J.; kleist, J. The relationship of drought frequency and duration to time scales. In Proceedings of the 8th Conference on Applied Climatology, Anaheim, CA, USA, 17–22 January 1993; pp. 179–184.

30. Lloyd-Hughes, B.; Saunders, M.A. A drought climatology for Europe. *Int. J. Climatol.* **2002**, *22*, 1571–1592. [CrossRef]

31.   Bordi, I.; Frigio, S.; Parenti, P.; Speranza, A.; Sutera, A. The analysis of the Standardized Precipitation Index in the Mediterranean area: Large-scale patterns. *Ann. Geofis.* **2001**, *44*, 965–978.

32.   Weedon, G.P.; Gomes, S.; Viterbo, P.; Shuttleworth, W.J.; Blyth, E.; Osterle, H.; Adam, J.C.; Bellouin, N.; Boucher, O.; Best, M. Creation of the WATCH forcing data and its use to assess global and regional reference crop evaporation over land during the twentieth century. *J. Hydrometeorol.* **2011**, *12*, 823–848. [CrossRef]

33.   Portmann, F.T.; Siebert, S.; Doll, P. MIRCA2000-Global monthly irrigated and rainfed crop areas around the year 2000: A new high-resolution data set for agricultural and hydrological modeling. *Glob. Biogeochem. Cycles* **2010**, *24*, 1–24. [CrossRef]

34.   Williams, J.R.; Jones, C.A.; Kiniry, J.R.; Spanel, D.A. The EPIC crop growth model. *Trans. ASAE* **1989**, *32*, 497–511. [CrossRef]

35.   Kamali, B.; Abbaspour, K.C.; Lehmann, A.; Wehrli, B.; Yang, H. Uncertainty-based auto-calibration for crop yield—The EPIC$^+$ procedure for a case study in Sub-Saharan Africa. *Eur. J. Agron.* **2018**, *93*, 57–72. [CrossRef]

36.   Abbaspour, K.C.; Rouholahnejad, E.; Vaghefi, S.; Srinivasan, R.; Yang, H.; Klove, B. A continental-scale hydrology and water quality model for Europe: Calibration and uncertainty of a high-resolution large-scale SWAT model. *J. Hydrol.* **2015**, *524*, 733–752. [CrossRef]

37.   Potopová, V.; Boroneanţ, C.; Boincean, B.; Soukup, J. Impact of agricultural drought on main crop yields in the Republic of Moldova. *Int. J. Climatol.* **2015**. [CrossRef]

38.   Brooks, N.; Adger, W.N.; Kelly, P.M. The determinants of vulnerability and adaptive capacity at the national level and the implications for adaptation. *Glob. Environ. Chang.-Hum. Policy Dimens.* **2005**, *15*, 151–163. [CrossRef]

39.   Damm, M. *Mapping Social-Ecological Vulnerability to Flooding-A Sub-National Approach for Germany*; Bonn, Rheinischen Friedrich-Wilhelms-Universität: Munich Germany, 2009.

40.   Kaufmann, D.; Kraay, A.; Mastruzzi, M. The Worldwide Governance Indicators: Methodology and Analytical Issues. *Hague J. Rule Law* **2015**, *3*, 220–246. [CrossRef]

41.   Tu, Y.K.; Kellett, M.; Clerehugh, V.; Gilthorpe, M.S. Problems of correlations between explanatory variables in multiple regression analyses in the dental literature. *Br. Dent. J.* **2005**, *199*, 457–461. [CrossRef]

42.   Eggen, M.; Ozdogan, M.; Zaitchik, B.; Ademe, D.; Foltz, J.; Simane, B. Vulnerability of sorghum production to extreme, sub-seasonal weather under climate change. *Environ. Res. Lett.* **2019**, *14*, 045005. [CrossRef]

43.   Montaud, J. *Agricultural Drought Impacts on Crops Sector and Adaptation Options in Mali: A Macroeconomic Computable General Equilibrium Analysis*; Revised 19; CATT - UPPA –University of Pau and Pays de l'Adour, 20 February 2019.

44.   Dumenu, W.K.; Obeng, E.A. Climate change and rural communities in Ghana: Social vulnerability, impacts, adaptations and policy implications. *Environ. Sci. Policy* **2016**, *55*, 208–217. [CrossRef]

45.   ASSAR. Understanding Vulnerability and Adaptation in Semi-Arid Areas in Botswana; 2015. Available online: http://www.assar.uct.ac.za/sites/default/files/image_tool/images/138/Info_briefs/Botswana%20Information%20Brief.pdf (accessed on 10 August 2019).

46.   Thornton, P.K.; Jones, P.G.; Ericksen, P.J.; Challinor, A.J. Agriculture and food systems in sub-Saharan Africa in a 4 °C+ world. *Philos. Trans. R. Soc.* **2011**, *369*, 117–136. [CrossRef]

47.   Zougmoré, R.B.; Partey, S.T.; Ouédraogo, M.; Torquebiau, E.; Campbell, B.M. Facing climate variability in sub-Saharan Africa: Analysis of climate-smart agriculture opportunities to manage climate-related risks. *Cah. Agric.* **2018**, *27*, 1–9. [CrossRef]

48.   Patt, A.; Gwata, C. Effective seasonal climate forecast applications: Examining constraints for subsistence farmers in Zimbabwe. *Glob. Environ. Chang.-Hum. Policy Dimens.* **2002**, *12*, 185–195. [CrossRef]

49.   Keshavarz, M.; Karami, E. Institutional adaptation to drought: The case of Fars Agricultural Organization. *J. Environ. Manag.* **2013**, *127*, 61–68. [CrossRef]

50.   Conway, D.; Schipper, E.L.F. Adaptation to climate change in Africa: Challenges and opportunities identified from Ethiopia. *Glob. Environ. Chang.-Hum. Policy Dimens.* **2011**, *21*, 227–237. [CrossRef]

51.   Simelton, E.; Fraser, E.D.G.; Termansen, M.; Forster, P.M.; Dougill, A.J. Typologies of crop-drought vulnerability: An empirical analysis of the socio-economic factors that influence the sensitivity and resilience to drought of three major food crops in China (1961–2001). *Environ. Sci. Policy* **2009**, *12*, 438–452. [CrossRef]

52.   Cochrane, L.; Bekele, Y.W. Contextualizing narratives of economic growth and navigating problematic data: Economic trends in Ethiopia (1999–2017). *Economics* **2018**, *6*, 64. [CrossRef]

53. Sandefur, J.; Glassman, A. The political economy of bad data: Evidence from African survey and administrative statistics. *J. Dev. Stud.* **2015**, *51*, 116–132. [CrossRef]
54. Jerven, M.; Johnston, D. Statistical tragedy in Africa? Evaluating the data base for African economic development. *J. Dev. Stud.* **2015**, *51*, 111–115. [CrossRef]

MDPI

St. Alban-Anlage 66

4052 Basel

Switzerland

Tel. +41 61 683 77 34

Fax +41 61 302 89 18

www.mdpi.com

Sustainability Editorial Office

E-mail: sustainability@mdpi.com

www.mdpi.com/journal/sustainability

Lightning Source UK Ltd.
Milton Keynes UK
UKHW050834190822
407497UK00002B/92